THE ROAD TO SCIENTIFIC SUCCESS

Inspiring Life Stories of Prominent Researchers

Volume 3

The Road to Scientific Success:
Inspiring Life Stories of Prominent Researchers

Series ISSN: 1793-2823

Series Editor: Deborah D L Chung
  (University at Buffalo, The State University of New York, USA)

Published

THE ROAD TO SCIENTIFIC SUCCESS

Inspiring Life Stories of Prominent Researchers

Volume 3

editor

Deborah D L Chung

University at Buffalo, The State University of New York

World Scientific

NEW JERSEY · LONDON · SINGAPORE · BEIJING · SHANGHAI · HONG KONG · TAIPEI · CHENNAI

Published by

World Scientific Publishing Co. Pte. Ltd.

5 Toh Tuck Link, Singapore 596224

USA office: 27 Warren Street, Suite 401-402, Hackensack, NJ 07601

UK office: 57 Shelton Street, Covent Garden, London WC2H 9HE

British Library Cataloguing-in-Publication Data
A catalogue record for this book is available from the British Library.

The Road to Scientific Success: Inspiring Life Stories of Prominent Researchers — Vol. 3
THE ROAD TO SCIENTIFIC SUCCESS
Inspiring Life Stories of Prominent Researchers

ISBN 978-981-124-726-2 (hardcover)
ISBN 978-981-124-757-6 (paperback)
ISBN 978-981-124-727-9 (ebook for institutions)
ISBN 978-981-124-728-6 (ebook for individuals)

For any available supplementary material, please visit
https://www.worldscientific.com/worldscibooks/10.1142/12550#t=suppl

Typeset by Stallion Press
Email: enquiries@stallionpress.com

Preface

This is Vol. 3 of the book series titled "The Road to Scientific Success". The series is aimed at inspiring people into careers in science, providing role models for people that aspire to pursue science, describing examples of pathways toward scientific success, reducing the hindrance to scientific success, exposing the readers to a wide range of scientific fields, providing examples of scientific success in various countries, and recording the history of science.

The book series is unique in its emphasis on the road to scientific success rather than the science itself. Scientists communicate much on their science through research publications, but they tend to talk much less, if at all, on the challenges encountered on the road to success. Information on the road to scientific success is helpful to people that are considering to embark on the journey on this road or are in the middle of the journey on this road. These people need inspiration and encouragement. Furthermore, unless the precious experience is recorded, it would be lost.

Four scientists are featured in Vol. 3. They are listed below.

1. Xiu-Yun Chuan – from extreme poverty to university professorship
2. Zulkhair Mansurov – from Kazakhstan to the world through combustion science
3. Merton C. Flemings – from America to the world through metal solidification science
4. Yiu-Wing Mai – from Hong Kong to the World through composites science

The touching and detailed account of Professor Chuan's incredible transformation from a village girl of extreme poverty in a cave dwelling in rural China to professorship in Peking University is a vivid illustration of the importance of the long-term determination to pursue one's dream (the scientist dream in this case) amidst extreme financial, cultural and life-sustaining difficulties. The story is worthy of being made into a movie.

In spite of the limited opportunities for scientific pursuit in Kazakhstan, Professor Mansurov's extraordinary dedication to scientific research and education over decades has resulted in his international leadership position in the field of combustion science. The learning of English, the loss of his wife and son, and the obtaining of research funding were some of the challenges that he faced. His firm focus on science enabled him to overcome the difficulties.

Professor Flemings of MIT is the world's most respected pioneer in the field of the solidification (casting) of metals. His love of science, his merging of science and applications, and his leadership in materials education are greatly cherished. He was the Head of the Department of Materials Science and Engineering of MIT when I pursued my Ph.D. degree there.

Professor Mai of University of Sydney, Australia, is an international leader in composites science. He was originally from British Hong Kong (currently China). His path in scientific education and research is rich, encompassing multiple countries and a variety of institutions. He applied his depth of knowledge in the mechanics of materials to composite materials, such as the carbon fiber composites that are used for aircraft. He has received high honors from Australia and China.

The life experience of Professors Chuan, Mansurov, Flemings and Mai is inspirational and historic. May the readers be envigored in their pursuit of science, in spite of hindrance.

D.D.L. Chung, Editor
Buffalo, NY, U.S.A.
October 17, 2022

Contents

Chapter 1

Xiu-Yun Chuan – from extreme poverty to university professorship

1.1 Introduction by the Editor

Born in a cave dwelling in a rural village in China, raised in a family that continuously fought against hunger and coldness, facing shortage of lighting for reading and paper for writing, having to contribute to farming and water transportation, and restricted by the tradition of women not needing to study, Xiu-Yun Chuan 传秀云 unbelievably overcame all these hurdles to pursue her dream since childhood of becoming a scientist. She studied in a university and persevered with inadequate fund to cover even her meals. In spite of having no funding or laboratory space in the beginning years of her university faculty career, she steadily climbed the academic ladder step by step and eventually became Professor of Mineral Materials Science in Peking University, one of the top universities in China. This path to scientific success from extreme poverty and the utmost degree of being underprivileged is extraordinary and breathtaking. Professor Chuan said that she was often in tears as she wrote her life story below.

1.2 Life experience as told by Professor Chuan

1.2.1 *Family in a primitive cave dwelling in the village*

1.2.1.1 *The struggles of my grandparents and parents*

When Japan invaded China in 1937 (a part of World War II), a tall man (my paternal grandfather, Mr. Bainian Liu 刘百年 (1887–1958) had to leave his hometown, which was about 10 miles from Lugou (Marco Polo) Bridge 卢沟桥 in Beijing. He escaped to Shaanxi (陕西省), where he lived as an uprooted person. There, he met and married a poor orphan girl (my paternal grandmother, Ms. Guihua Liu 刘桂花, 1907–1962), who had lost her parents during the War. They started a family with two other poor girls and one boy, who had lost their father in the War. They raised their children in a small village (Xuzuoxicun, 徐卓西村) in the countryside in northwest China.

The village of Xicun (西村, abbreviation for Xuzuoxicun) is in the southern part of the Loess Plateau (黄土高原, literally meaning "The Yellow Earth Plateau", Fig. 1.1) in northern Shaanxi (陕西), specifically in the north of Chengcheng County (澄城县, located north of Weinan City 渭南市). The village is 4 km long from east to west, with an area of about 20 square kilometers. It is one of five villages in Xuzuo (徐卓).

The region covers almost all of the northern provinces of Shaanxi, Shanxi, etc. Homes called yaodong (窑洞) were often carved into the loess (German meaning loose) soil in accordance with the complicated terrain

Fig. 1.1 The Loess Plateau (brown shaded region) occupies an area of 640,000 km^2 (250,000 sq mi) in northwest China. Its elevation is 1,250–2,000 m (4,100–6,560 ft). The Yellow River is colored blue. The yellow area is Inner Mongolia and Ningxia. https://en.wikipedia.org/wiki/Loess_Plateau (Public domain)

of the area, in order to provide thermally insulated and low-cost shelter, as needed for the cold winter and hot summer. Our home was a cliffside yaodong (靠崖窑), which was built by digging a cave in the cliff. Such underground dwellings dated back to the 2nd millennium BC (the Bronze Age and the Xia Dynasty 夏). The loess soil is highly erodible, having been deposited by windstorms that commonly occurred over the ages. Due to the erosion, the hilly landform was full of deep gullies.

When World War II and the immediately subsequent War of Liberation ended in China in 1949, peaceful life seemed to be on its way to my paternal grandparents. However, my paternal grandfather and grandmother died in 1958 and 1962, respectively, after the marriage of their two elder daughters. Thus, they left behind their youngest daughter and only son, my father, Mr. Chun Chuan 传春 (1943-), along with his poor wife, my mother, Ms. Zhulan Wang 王竹兰 (1945-).

My mother lost her mother (Guixiang Wang, 王桂香, 1927–1945) when she was only several months old. She was raised by her poor father, Mr. Shenshun Wang 王生顺 (1918–1991) and her elderly grandmother. My maternal grandfather was a diligent man adopted by a kind couple as a begging orphan ever since he was about 7 years old (during World War II). He earned money by selling homemade soft glutinous rice cake mixed with red jujube, like traditional Chinese rice-pudding without packing, in the small town (Fengyuan 冯原), close to my village (Xicun).

My parents made great efforts for their lives alone and had to help in the marriage of my father's younger sister (my aunt Chunlin Chuan 传春莲). Then, I was born in the mid-1960s as their second daughter.

1.2.1.2 *My first taste of culture*

I was born and raised in a peasant family in a poor village in the midst of extreme poverty. However, I was fortunate to have grown up under the influence of the traditional cherished Chinese heritage that encouraged culture, particularly cultivation (quality development) and reading. My small village is about 200 km from Xi'an city (西安), but close to the old tomb built before 162 A.D. (Eastern Han dynasty, 25–220 AD), and less than 10 km from the old Memorial Temple of more than 1800 years of Cangjie (仓颉), the father of the Chinese characters (Fig. 1.2). Cangjie created the Chinese words as pictographs on the basis of the different images and the

Fig. 1.2   Cangjie (仓颉), a legendary figure with four eyes in ancient China (c. 2650 BCE), supposed to be an official historian of the Yellow Emperor (黄帝) and the inventor of Chinese characters. https://en.wikipedia.org/wiki/Cangjie, public domain.

Fig. 1.3   Canjie devising Chinese characters with the inspiration he obtained from the birds.

moving traces of the birds, animals and other objects (Fig. 1.3). Instead of the use of knots (akin to Quipu of Andean South America) to record information, he devised a writing system in the form of characters (Fig. 1.4). He did this work when he was a lowly historiographer at Shiguan village (史官), about 8 km from my hometown. Due to the short physical distance, we were able to visit his working site at Shiguan village, his living place at Wuzhuo village (武庄), in addition to enjoying the stories about his cleverness, as told by the adults and teachers. These stories enlightened me and inspired me to be clever like Cangjie, even in my childhood. For many families with children, it was a part of the traditional culture to visit the tomb of Cangjie and pray in his temple in the Chinese (Lunar) New Year every year (Fig. 1.5).

1.2.1.3 *My home in a primitive cave dwelling*

Cave dwellings with small gardens (some of them at the rooftop) were common in the 1960s in the Loess Plateau village where I grew up

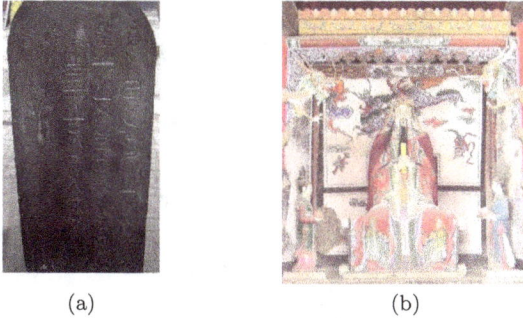

(a)　　　　　　　　　　　　　　(b)

Fig. 1.4　(left) Chinese characters devised by Canjie. (right) Canjie's temple.

(a)　　　　　　　　　　　　　　(b)

Fig. 1.5　The family visiting in 2006 the 2000-year old Memorial Temple (a) of Cangjie 仓颉, a legendary figure in ancient China (c. 2650 BCE), supposed to be an official historian of the Yellow Emperor (黄帝, reign in 2698–2598 BCE (mythical)) and the inventor of Chinese characters. In (b), the family members are adults at back, including my father Chun Chuan 传春 (back row, left 1), mother Zhulan Wang 王竹兰 (back row, left 2), me (back row, right 2), my husband Jianhai Li 李建海 (back row, right 1), and children at the front, i.e., my niece Tian Chuan 传甜 (front left), my nephew Hao Chuan 传昊 (front center), my son Chuanjin Li 李传津 (front right).

(Fig. 1.6(a)). Some dwellings had heavy stone doors. This was particularly common during World War II, because a stone door was heavy and could not be moved easily from the outside. Thus, the heavy sandstone door of about 10 cm in thickness was useful for protecting the family. However, such doors have been removed by now. People used simple farm tools, such as the big cylindrical stone, Luqiao (碌碡) (Fig. 1.6(b)), which is a type of

(a)   (b)

Fig. 1.6 Examples of cave dwellings (a) and the heavy sandstone door of about 10 cm in thickness (b) in the hilly Loess Plateau.

old farm tool used to roll the grain, wheat and bean, on a flat field. In spite of the hardship, the people led simple peaceful lives there.

I was born in one of the cave dwellings and lived with my parents and three siblings (a younger brother, an elder sister and a younger sister). It was unusual to have photos taken at that time. Thus, I was lucky to have the first image of mine, also the first for my elder sister (Yayun Chuan 传亚云) and younger brother (Jianwu Chuan 传建武), particularly the first image of my father with his children. Unfortunately, my mother was not in the photo, which was taken in 1969 (Fig. 1.7).

Usually a family used two caves, with each cave serving as a room. One of the caves was for both the bedroom and the kitchen. The other cave was for living and storage. Only one wide heatable adobe earth platform, made of clay and wheat straw, could be used as the bed for the whole family, including the parents and their children, as no wooden bed could typically be afforded. However, a simple wooden bed consisting of several wood boards (about 1.5 meter long) supported by two long wooden stools was used by my brother when he became a teenager. In the summer, the living area located near the gate of the cave was sometimes separate from the storage area. This separation was possible for families that were relatively rich (Fig. 1.8).

There was not really a kitchen in the cave for my family. Beside the wide heatable adobe earth platform for the family to sleep on, there was a small area with a self-assembled stove for cooking (Fig. 1.8(a)), along

Fig. 1.7 My first image (front right), with my father (back), younger brother (front center), and elder sister (front left), taken in 1969.

with a cooking board in the form of a wooden plank supported by two long wooden stools (Fig. 1.8(b)). My mother prepared vegetable, noodle, buns and almost all types of food on this chopping board (Fig. 1.8). Particularly memorable are the beautifully decorated Chinese steamed buns (mantou 馒头) for ceremonial celebrations, such as birthdays, weddings, funeral and festivals (Fig. 1.8(b) and Fig. 1.8(c)). The dough is specially shaped and decorated to resemble animals (e.g., chicken and birds), flowers and other objects. The beautiful buns are known as Huamo (花馍). They are used as gifts for festivals (Fig. 1.8(b) and Fig. 1.8(c)), birthdays (Fig. 1.8(d)), and funerals (Fig. 1.8(e) and Fig. 1.8(f)). They represent a part of the folk art of Shanxi. As a child, I was always very excited when I saw the beautiful buns when they came out of the kitchen upon completion of the steaming over the kitchen fire. They were both artistic and funny. I enjoyed the buns so much that I tried to copy them and create new styles by using loess mud instead of dough. Some neighbors even invited me to help them in the preparation of their buns for special occasions.

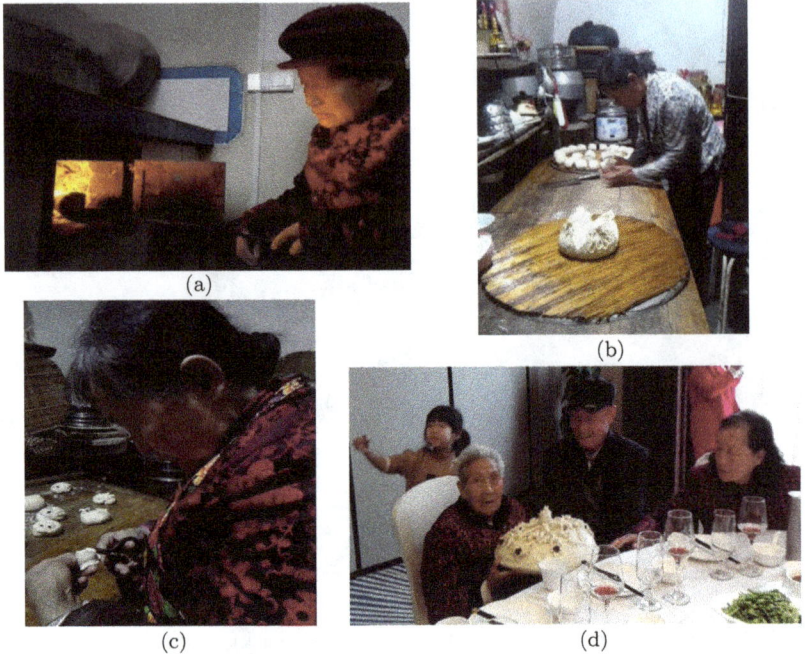

Fig. 1.8   My mother preparing Chinese ceremony steamed buns using dough in the shape of animals in 2018. She used a chopping board (supported by two long wooden stools) and a kitchen stove located in the village courtyard in the hilly Loess plateau. The photo in (d) shows a Huamo gift given in 2018 by my father (front center) and my mother (front right) to my 90-year old aunt (front left), the wife of my father's brother, with her great granddaughter (back). The Huamo gifts, (e) (top view) and (f) (side view), used in the funeral of an old relative in 2000.

I greatly appreciate my mother's unusual skill in making by hand beautifully embroidered objects in the form of cats, shoes, etc. (Fig. 1.9). I also enjoyed to model after her artistic style in preparing the shoes myself. The work gave me a sense of achievement and strengthened my interest to create new beautiful styles by using different lines, circles and colors. This contributed to cultivating my interest in science ever since my childhood.

1.2.1.4 *Playing with homemade toys*

There was typically no commercial toy for the children to play with in the village. Therefore, I made the toys with the help of my parents. My mother

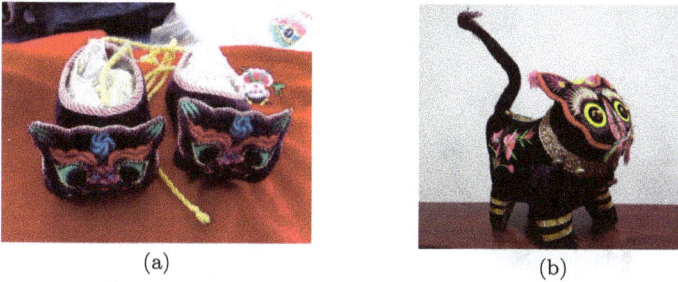

(a)   (b)

Fig. 1.9   Hand-embroidered shoes (a) and cat (b) prepared by my mother in the village courtyards in the hilly Loess plateau around 1998.

helped me to find some pieces of old cloth, and taught me to sew the cubic sandbags (Fig. 1.10(a)) manually. I played the game with other children by throwing a small sandbag at each other (Fig. 1.10(b)). I admired those boys who could run the steel rings (Fig. 1.10(c)) with the long hook, as made manually by their parents. The shuttlecock (Fig. 1.10(d)) was a popular toy made of the chicken feather and an old coin. We often played by kicking the shuttlecock in order to keep warm in the winter. Sometimes we had fun to find a rope from the trash to jump over.

Chinese chess (象棋) was a game that I enjoyed. We used small stones, waste wood branches and a makeshift chess board drawn on the soft loess ground, as the adults did this often. This game also helped me think and strategize.

I greatly enjoyed playing with the other children. We climbed up the steep slope, ran down fast from the loess dam, and wandered everywhere in the village. It was fun to play the game of hide-and-seek, but it usually took too much time. Sometimes, the game was over and the other friends had disappeared, while I was still hidden somewhere.

The games without nice toys may seem pointless, but they actually provided training for the body, athletics, mind and patience. I treasure the sweet memories, in spite of the poverty.

1.2.1.5 *Obtaining water from a well by using a pulley*

The homes in the village were very simple and primitive. Due to the dry season in the loess plateau, water was precious and had to be obtained from

Fig. 1.10   Homemade toys of the children, sandbags (a, b), steel rings (c), and shuttlecock (d), in the village.

the village well of depth about 60 meters. There was no water pump or water pipe. Potable water had to be obtained weekly from the well by using a block and tackle pulley system (Fig. 1.11(a)) along with a big bucket tied by a long rope. Two buckets, each filled with water, were then carried by using a long pole supported at a shoulder (Fig. 1.11(b)) from the well to a big black ceramic water tank that was about 100 cm in diameter (a traditional black Chinese ceramic product made in Chengcheng 澄城 county, Weinan 渭南 district, Shaanxi 陝西, China). This hard work for the whole family was the most important duty of the stronger members of the family, such as the father and brother, Similarly, in the harvest season, grains needed to be stored for feeding the whole family for the months ahead.

In the busy season for the crop sowing or harvest, all people had to participate in the labor. Thus, my elder sister and I often had to carry two

(a)   (b)

Fig. 1.11   Potable water obtained from the well by using a block and tackle pulley system (a) along with a big bucket tied by a long rope, (b) Carrying two buckets filled with water using a long pole supported by a shoulder.

buckets of water with a long pole at the shoulder. However, in most cases, the water was partly spilled from the buckets onto the road due to my inadequate balancing. Therefore, I resorted to filling the buckets at a little more than half full, and took more time to transport the water carefully (Fig. 1.11). Over the years, the water supply system had been improved greatly, but the big black ceramic tanks were still kept by many families and were often used, as the water supply was often not on time. However, since the 1990s, water was supplied from deep wells drilled mechanically, using water pipes and pumps. Some of big black tanks have still been used to store the harvested crops, such as wheat, corn and bean. Hundreds of old big black tanks have been collected and kept as the historical heritage with the primitive earth kiln in recent years.

Water from water pipe could not be used to clean the lavatory, which was about 50 meters from the cave. In addition, new loess had to be collected and recycled often to cover and absorb the dust, and also for the pig manure cleaning. As the oldest children, my elder sister and I had to help our parents clean the lavatory and pigpen. This also involved applying fertilizer earth to the plants on the ground by using a heavy wood cart. It was hard work for young girls to carry heavy carts filled with loess from the deep slope. This induced my interest to try to look for a better method. I was so excited to find ingenious ideas, such as the mechanical theory associated with the slope, the pulley system, the lever principle, and the electromagnetic field in physical textbooks and classes. This inspired my interest in physics very much. As a consequence, I often received the highest score in my high school, though it was unusual for a girl.

1.2.1.6 *Being industrious*

My mother often remarked, "Be an industrious person", "Anybody can be appraised and become valuable through diligent labor", "There isn't any useless effort in the world". The labor is important not only for physical achievement, such as agriculture, but also for one's intelligence. It is beneficial to everyone, including children, youths and adults, whatever is their form of labor. Considerable hard physical labor is involved in planting crops in the ground and in housework. Due to the critical need of physical labor for sustaining a family, men were often considered more important than women, at least in the underdeveloped countryside. This notion was so widely accepted that men rather than women inherited the family fortune and the family name, thus causing women and girls to be subordinate, with little chance of advancing in education or career.

1.2.1.7 *Studying without adequate lighting*

No electric lighting was available to me till I was in junior middle school. Only kerosene lamps were available for my use in studying in the evening (Fig. 1.12). Due to the soot, my noses often became black inside. There was no public lamp in the classroom, so we had to prepare the lighting ourselves and bring the lamp to school. With only one lamp in my family, I could not take it to school. Otherwise, no lamp would be available at home. Furthermore, my elder sister needed it in her high school classroom too. Therefore, I often had to share the kerosene lamp with some classmates, though this sometimes resulted in humiliation by some classmates, particularly some mischievous boys.

The glassy shade was sometime used for the lamps in the classroom, but it was not used at home, due to its cost. I often had to do exercise and homework by sitting at the edge of the bed and using the only lamp (without a glassy shade) at home (Fig. 1.12(b)). Another difficulty was the frequent complaint by my step grandmother of my seemingly excessive use of kerosene. I had to live at her home when I attended middle school. Due

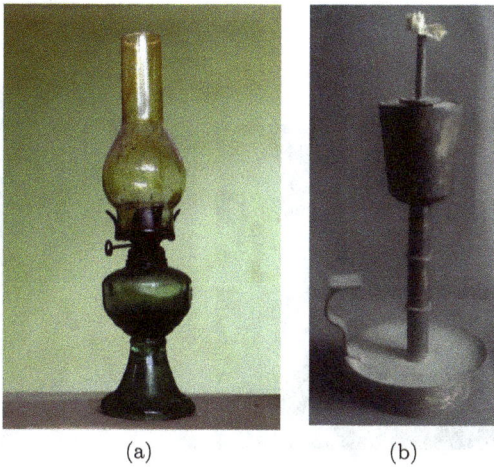

(a)   (b)

Fig. 1.12   Kerosene lamp with glassy shade in the classroom (a) and without the shade for the lamp at home in the village (b).

to this situation, I often read and did homework by using the light coming out from the window of my teacher. I sometimes even worked under the moon light. In order to make the most of the time when light was available, I usually read books on the road. Unfortunately, this often caused my hitting trees and the need to say sorry. I also made use of the lunch time (when natural light was available) to read books and recite texts. For this purpose, my lunch was extremely simple. Fortunately, I was satisfied with simple meals, such as one consisting of a bun made of bran wheat along with only salt and hot pepper that had been mixed with water. That was my life for at least several years when I was a child.

It was very expensive (even extravagant) to support every child in a family to go to school. I was fortunate to have the chance of going to school, in spite of the need for expensive kerosene at night.

My parents asked us to get up early, as they did often for farm work. They often got up at sunrise and went to bed at sundown. This practice was economical and efficient for me as well, as I could make good use of the time of free natural lighting to read books and study.

There were about 20 families in the village. They made up a population of about 70 people. When I was a primary school student, there were only 6 classmates at my age. The two boys among these six classmates stopped

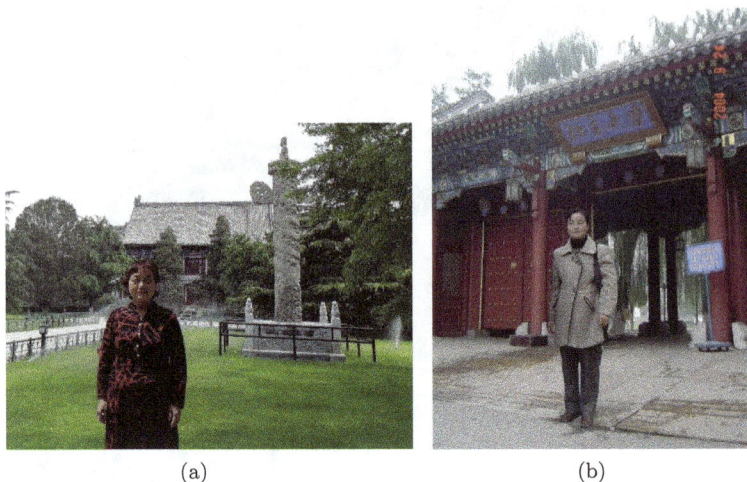

(a)   (b)

Fig. 1.13   Peking University campus in 2021(a), and the traditional west gate (b) in 2004 Beijing, China.

their schooling in their primary school years, the other two girls among the six left school for marriage at the first year of their middle school. One other girl among these six went on to attend a university, but she soon ended her education in order to work in a factory for the sake of earning a salary. I was really fortunate to be able to go to school and then university, where I received the master and doctor degrees. It was a miracle. When I later became a professor in a top university (Fig. 1.13(a)) of China, namely Peking University (北京大学) (Fig. 1.13(b)), I became the symbol of the village.

1.2.1.8 *Food shortage in the village*

There was only one way to earn a living in the village. It was to plant the crops, such as wheat, corn and beans, for consumption by the family. However, the production quantity was very limited, because of the harsh climate and scant rain. Many families were short of food by the late winter and early spring of each year. Typically, there were only two dishes on the dining table, without any vegetable. One dish consisted of rocky salt particles, while the other dish consisted of chilly power mixed with hot water (or oil for those that are relatively rich). A small piece of meat was on the dining table only once a year on the Lunar New Year's Eve. There was one exception, which pertains to the meat used in a funeral ceremony. Although this meat was on the dining table on the occasion of funerals, I was awfully scared for a long time as a little girl, thinking that it might have come from the dead.

Usually no food was allowed to be brought to school for a child's lunch. All of the students had to return home for lunch, which typically consisted of only soup that had no meat or protein. Both the quality and quantity of the food was not enough for active children. This was how I lived till my high school days. Once my younger sister complained about her lunch when she was in her middle school. Her lunch consisted of only a small bowl of soup and streamed buns made with a mixture of wheat and bran. My neighbor then told her that there was no lunch at all for my mother.

The poor situation persisted till my high school days, even though my parents worked very hard and we contributed by looking for wild herbs and helping with the cooking, and picking up wood branches for boiling water. Our help was particularly needed when my parents were busy.

Life improved a little during late spring and hot summer due to the availability of wild herbs and leaves, whether edible or not. Life improved further after the autumn harvest. I still remember the absence of grain or vegetable when my mother was ready to cook with boiling water. My mother shed tears in front of her four hungry children. She usually took some dry wild herbs in the cold winter and fresh leaves in summer, often regardless of whether they were edible or not. Some of the leaves were from the tall trees, Ailanthus altissima (臭椿), not Toona sinensis (Chinese mahogany 香椿), that were planted using the seeds that my paternal grandfather brought from his hometown in Beijing.

Later, life improved after the importing of the sweet potato plant to the village. However, the sweet potato became an edible food ingredient (often mixed with wheat bran or corn) only when I was in high school. The poor quality of the food resulted in serious constipation, particularly for many of the children, most of whom were scared about the food, even two decades later.

Life was difficult for everyone, so help was not available from any relative, including my maternal grandfather and my maternal step-grandmother. My grandfather had to take care of his four hungry children (the younger siblings of my mother). Everyone had to take care of oneself firstly in that hard period.

1.2.1.9 *Illness with help from the witch and doctor in the village*

There was no healthcare service in the village. My maternal grandmother died when she was less than 30 years old. This was when my mother was less than 1 year old. This was before 1949, which marked the end of the Chinese Communist Revolution, known in the mainland China as the War of Liberation. The Revolution resulted in the proclamation of the People's Republic of China. My paternal grandmother had a serious stomach illness, which caused him to be unable to eat anything. She died in 1962 when she was 55 years old, although my father took her to a hospital that was 30 km from the village.

Some villagers were trained as the village doctor after the War of Liberation in China in 1949, but typically most families did not have enough money to see a doctor. As a consequence, clues were sought from the witch, who performed a special ceremony, in which people danced with their spears and "sang" to their "god". Sometimes a drink for the sick was prepared. This drink consisted of water from a designated river and the ash of burnt paper with words of blessing written on the paper. It is not easy to accept this practice, but the villagers did whatever they could to help their relatives recover.

In order to get a bottle of the special water for me, my father once took a day to go to a deep ravine at a distance of 8 km from the village, to find the small river according to the instruction of the witch. I have forgotten about the efficacy of the special water, but I greatly appreciate how much my parents cared for me, just their second daughter.

No one in the village could live to an old age, due to the poor health care service and bad economic situation. My maternal grandfather died after a small cold in 1991, my maternal step-grandmother in 2000 and my two aunts died after small leg fractures. My father fainted while he taught in the classroom in the 1980s, but he did not go to a hospital, due to the bad economic condition.

The situation improved after a small hospital had been built. That was when I was sick in high school. I remember my parents taking me to see a doctor in the hospital. I was lucky not to have experienced any serious illness as a child, but I did use the special water from the witch at times.

Now the villagers can live much longer than before. With the improved health care service, I am now able to see and stay with my parents. Now I willingly pay for the hospital service when my parents are sick, in spite of the fact that the health insurance pay some for the service. I remain worried of the chance that I might lose my dear mother due to a small illness, as I do not want to become a daughter without a mother. The endless tears of my mother still stay in my mind vividly.

1.2.1.10 *Reconditioning clothing every season*

Life was so difficult that one felt lucky to be alive. Due to the hard work of my diligent parents, my siblings and I were fortunate to have thin but clean clothing during the cold winter. My mother had to modify our scarce clothing whenever the season changed. She did so by removing the cotton fill for the warm spring, removing the lining for the hot summer, and reinstalling the cotton fill for the cold winter every year. I often got the old clothing from my elder sister and my clothing was also often given to my younger siblings when I had outgrown it. Nevertheless, I was much more fortunate than my parents during the 1950s, when there was not enough cotton clothing for everyone and the plastic bags used to contain fertilizers were used to make outdoor clothing, such as trousers made after dyeing the bag material black. Particularly for the Spring Festival (Chinese New Year), "new" clothes were made by hand for me and my siblings by my parents, using cloth fragments obtained from the outgrown clothes of older siblings and selected parts of the clothes of some adults. It was popular for many adults, such as my parents, to wear the cloth frugally in the hard duration. One cloth used to be worn for three years as the new only in the ceremony, for three years as the old to use in normal life and another three years as the mended to wear in the farming.

A boy in the village nearby was much less fortunate than me. He was an orphan, because his mother left him after his father died. All of his clothes came from kind villagers, though the clothing did not always come on time. As a result of inadequate warm clothing during the cold winter, he kept warm by moving strangely in the middle school and thus became a joke. I was very pleased to know that he went on to become a university student.

1.2.1.11 *Burning fire pan for warmth*

In every cold winter when I was child, I suffered from deep splits in and below the skin of my hands and feet. They were very painful.

It was very cold in the winter, not only outdoor, but also in the classroom. There was no heating system in all of the schools in the village. Following the school arrangement, I kept warm by running for about half an hour in the early morning before class.

Like the other primary school students, I often took a small burning fire pan to keep warm as I walked on the road to school (Fig. 1.14). In addition, I put the pan under the desk that was made of a composite of clay and plant straw to help keep me warm in the classroom. It was not easy to keep the fire burning as one walked on the road, so I often engaged in discussion with my friends about the technique of lighting the fire and maintaining the

Fig. 1.14 Carrying a small burning fire pan for warmth on a road in the village courtyards in the hilly Loess plateau.

sparks for a long time. Thus, I became known as a skillful lighter among my primary school friends.

When I was in middle school, my friends and I often ran, skipped a rope on one foot and played various games that involved physical exercise in order to feel warm. Concentrating on the gaining of knowledge also helped us forget about feeling cold.

1.2.1.12 *Kindergarten and primary school in cave dwelling*

When I was about 7 years old, I was fortunate to attend the first kindergarten in the village. The school was in a simple tile-roofed house, which are used for the bridal chamber of the adult son of the kindergarten founder. It was also used as the living room for the main cave dwelling in the village courtyard. A year later, at the age of 8, I started to attend the primary school in the village along with all children at the same age in the village. The kindergarten was closed after a year of operation and there had been no kindergarten in the village till recently. Some of children attended a private kindergarten outside the village but close to the town.

In the primary school, all the teachers were strict and knowledgeable men, like my father. However, the facility was very poor in the primary school. No clock was available, so the time was announced by the ringing of a bell tied to a big tree using a rope by one of the teachers. My schoolmates and I went to school in the early morning during sunrise. After having some classes in the school, we went back home for lunch at noon, when the shadow of the sun indicated that it was midday. After lunch, we returned to the school to have other classes in the afternoon. Upon finishing the day of classes, we returned home during sunset. We did so on every weekday, including Saturday. We often took a rest on Sunday, though Sunday was usually spent on homework, housework and field work. After 5 years, I graduated from the primary school.

The primary school was about 2 km from my home, so it was not easy to go there on foot. In my spare time, I enjoyed looking for the different footprints made by passersby and making footprints using my shoes on sunny days, as enabled by the flexible surface of the road. However, it was very hard to arrive at the school on time on the rainy days, due to the soft loess road and my inadequate shoes that were made of old cloth fragments.

Thus, I started to dream about a flat road with road lamps for the village someday. Sometimes I was scared to walk, even during my middle school years, as the school was about 4 km from my home and I had to return home when it was dark. It was particularly scary when it rained and there was no moonlight. To help me relax, I often tried to recite texts, such as poems about the moonlit night. This habit turned out to help me remember and understand the Chinese texts, in addition to improving my Chinese language ability.

1.2.1.13 *School tuition in the village*

School tuition in the village was not high, normally 2 Yuan RMB per semester for a primary school student, 8 Yuan RMB per semester for a middle and high school student in 1970s and 1980s. However, it was too high to be afforded by many peasant farming families. Only a few students could get free tuition, such as orphans, like the poor boy mentioned above. For a typical peasant farmer, as for my mother, it was quite difficult to receive income in the form of cash, besides the grain harvested.

We are proud of my father's salary as a primary school teacher. However, actually, I often got into trouble, due to my late payment of my tuition, particularly in middle school. The tuition of my three siblings and mine had to be paid at the same time at the beginning of every semester. It was very difficult for my father, due to his low salary, about 4 Yuan RMB per month. We typically had to postpone the tuition payment, sometimes till the last month of the semester, using the salary of my father received in the last month of the semester. Actually, my parents were often helpless, not being able to pay the tuition of their four children at the same time. Nothing was left from my father's salary after our tuition payment. However, my parents did not mind that, as they viewed it important for all of their children to attend school.

As a stubborn, positive and active teenage girl in middle school, I was ashamed of my late tuition payment. I often waited for the money for the tuition by leaning against the entrance of the school for a long time. Sometimes a colleague and friend of my father gave a helping hand and lent some money to my father. In particular, this happened to cover the cost of my university entrance examination.

Without any other choice, my father once had to ask me while I was a primary school student to try to ask his peasant friend in a nearby village for the repayment, that seemed to be an interest-free cash loan that my father gave to his friend to cover the tuition of his child in his hard time before. I went there alone, but I failed to obtain any money after waiting for a day at the front of their cave dwelling. I came back in hunger and hopelessness, with tears, though it was understandable for every family in that hard time.

There were some seedlings and sapling grown naturally from the seeds of Ailanthus Altissima and Tung tree in the court. In order to get enough money for the tuition earlier, once my father asked me to try to dig out the seedlings and sapling and sell them in the rural fair market. In advance, I calculated the price and the numbers of seedlings and sapling for numerous times in my mind. I took the seedlings and sapling on my shoulder in the early morning of a fair day, but only a few were sold till the end of the fair at sundown. I was very regretful for losing the first buyer, who wanted to buy, but only at a low price. The price might not be low, but it was not enough for my tuition. The price came down continuously during the day, and I became disappointed, regretfully being unable to sell much. I tried to analyze what went wrong and concluded that I was too eager to sell enough to cover my tuition. I should have been more patient.

1.2.1.14 *School life in the village*

We had classes on various subjects, such as mathematics, the Chinese language, painting, sports and music. The teachers were nice, knowledgeable and strict. They knew much about the method of creating the Chinese characters as used by Cang Jie, and also about Huanghe (Yellow River 黄河), the longest river in China, only 200 km from my village. These subjects attracted me very much.

The condition of the school was very poor. There were only five large caves for use as classrooms. The desks were made by hand totally from loess mud and wheat straw. In addition, there were five small caves for use by the teachers as the office and dormitory.

There was also a loess square without any sports facilities, but we were happy to run back and forth between the two ends of the square in games titled "the dog to catch the rabbit" and "circle running to spread the small handkerchief". I was usually afraid to take part in the sports with my 30 classmates in the primary school, due to my shoe sole not being well connected with the upper part of the shoe when I was running.

1.2.1.15 *Making shoes*

I hardly had a pair of shoes without hole and with heels during my entire childhood. The situation was essentially the same for the other children in the village. All my shoes were sewn manually by my mother, who made the entirety of every shoe from the sole to the upper part by using layers of cloth fragments joined by a soft paste and knitted together using cotton thread. Though the making of the shoes was difficult, the shoes did not last long, particularly since I was an active child. My elder sister and I were asked to learn the technique of making shoes. As teenagers, we spent time to help my mother knit the shoe soles. Many of the girls at that time did this. Even now, some of the women still need to make shoes for their parents.

Shoe making was boring and time consuming, but it gave me an opportunity to practice knitting and sewing. This experience turned out to help me with my research on making composite materials much later, when I became a university professor.

1.2.1.16 *An English letter from the U.S. and the study of English*

I did not learn any language other than Chinese when I was in primary school. However, one day, English became exciting to learn. This happened in 1978 when I was in the second semester of middle school. An English letter came from the U.S. It was written by a Taiwan person for her son, who was one of my mathematics teachers in the school. She was born in the neighboring village and had to leave her hometown with the Kuomintang (Chinese Nationalist Party 中国国民党) army during the War of Liberation in the 1940s. She had not come back to see her son for many years. Therefore, she was assumed to be dead and offered the sacrifice for the dead in her hometown every year.

Nobody in the village could Figure out the meaning of the letter, so the teachers at Xi'an Foreign Language University were consulted. Fortunately, an English course was started in my middle school in that year. My schoolmates and I were very proud of our English teacher, who could understand a part of the letter. However, he told us that it was difficult to understand the full meaning even with the help of an English dictionary. He encouraged all of his students to study English hard. Due to his help with reading the letter, we respected him as our English teacher. Actually, he had been trained somewhere in the teaching of English only for several months.

I started to learn English with strong interest in middle school due to the letter. In high school, I studied hard with the help of another English teacher, who was in charge of my class after he had received his bachelor's degree from Dali Normal College (大荔师范学院). His English knowledge was superior to the textbook. I became the class representative in the English class and became increasingly interested in studying English. This interest continued after I became a university student.

1.2.1.17 *The conflict of school study and farm labor in the village*

Due to the hard physical labor associated with farming, it was impossible for the parents to spend much time in helping their children grow up and study. Most of the families stopped their children's schooling at some point. This was particularly common for girls, who were not considered family member after their marriage.

Many of the parents who grew up before the end of the War of Liberation in 1949 were illiterate. Though they agreed to send their children to school, they did not mind stopping their schooling once they were able to write their names, perform simple calculations and recognize the money amount, particularly in the case that they needed help with the physical labor associated with the farm, which would be a family inheritance for the children in the future. For example, my younger brother had difficulty continuing his study after some absence from classes due to a curable illness. Compared with being alive, education was not important, even for boys. This was particularly common for the parents who had lost their fathers or brothers during World War II and the War of Liberation. Most of the boys of my age, or even younger, stopped their study after middle school, and became farmers. They remain as farmers in the village even now.

With a small primary school inside my village, the situation was much better than other villages. In my village, all the children finished their primary school education and went on to middle school, but most of them did not complete middle school, even more of them did not complete their high school. Only a few could go to university. The situation has improved now, of course, but the population has been continuously decreasing in my village. This is unfortunate.

I commonly experienced starvation and my clothing was shared, old and inadequate. As a kid, I participated in much labor at home, such as taking the grass for feeding pigs, picking up the feces of the sheep for fertilizing the crops, collecting the lost wheat seeds in the school, cooking, making shoe pads, sewing shoes, enriching the soil with fertilizers and feces, carrying water from the deep well, and so on. I fortunately grew up in spite of the limited care from my hardworking parents in a period of great difficulty in China.

1.2.1.18 *Unique girl and her mother in the village*

I was really a unique girl in the village. This is probably because of my mother, who was born in the 1940s. Regrettably, she lost her dream of completing her studies in a school, because of hard housework and poverty and the loss of her mother in her childhood. Therefore, her hope rested on her daughters.

Knowing that I was an active child, my mother allowed me to go to school even in the midst of troubles. In one instance, a lady complained to my mother that the skin of her daughter was wounded when her daughter and I climbed a tall tree together in order to pick some wild fruits. Mother did her best to provide for her children to the best of her ability. She often raised chicken for the eggs, which she sold to pay for the tuition and study fees of her children. However, she did not eat any of the eggs. She prepared many kinds of delicious food using wild herbs, wheat bran, potato and corn, and bravely tried to eat the leaves which had never been eaten by anyone, so as to save us from starvation in times of difficulty. Most ladies stopped sending their daughters to school due to financial difficulty and the influence of tradition, and asked the daughters to learn the skill of fine needlework and to make shoes for the family members. Some of the ladies advised my mother to stop sending me to school. However, mother bluntly refused without hesitation Even my aunts gave similar comments to my mothers, particularly when my mother borrowed money from them for paying the tuition of her children in times of hardship and hunger. My mother's answer was clear, "My daughter can buy the shoes in case that she cannot make them well". Nevertheless, my mother taught me how to do needlework and cook, and she did not push me to learn more about sewing, weaving and flower bun making. However, she prepared a lot of shoe soles for family members, including me and my husband, every year. She does so even now, when I can afford to buy commercial shoes.

Except for the hunger, I was happy to have grown up in a village as an underprivileged girl. *"Due to my low social status, I had been engaged in many lower class jobs when I was young, so I mastered a lot of skills and techniques"* (吾少也贱，故多能鄙事) written in the Analects (论语) by Confucius (孔子, 551–479 BC, The founder of Confucianism, one of ancient china's most famous ideologists and educators). I am grateful for my parents, who had open minds and made it possible for me to go to school and search for my scientist dream in the university.

1.2.2 *A little girl with the scientist dream*

1.2.2.1 *Never leaving my hometown until I was 18 years old*

It was customary in the countryside for allowing only boys to attend family activities, such as marriage and funeral. I understood this kind of arrangement, but my parents tried to take their daughters to family activities, in spite of the criticisms given to them by the village elders. Such criticisms were particularly common in times of food shortage. Only unpaid waitresses were allowed, though nobody took care of their little daughters, whether the girls were hungry or not. This situation caused me to be scared of family activities. In addition, I was awfully scared about the origin of the rarely seen meat in funeral ceremonies.

Due to the underprivileged situation of the girls in the distant countryside of China, I never left my hometown till I was 18 years old in 1983, when I became a university student. However, leaving was very difficult, mainly due to the travel cost.

Fig. 1.15   In 2016, after about two decades as a professor in Peking University.

1.2.2.2 *My entering a key university of China surprising my village*

My entering a key university Changchun College of Geology (presently College of Earth Sciences, Jilin University 吉林大学) was a huge surprise for my village. My later becoming a professor in Peking University (北京大学) was an unbelievable miracle.

By now, I have been a professor for about two decades (Fig. 1.15). However, the spirit of an indomitable girl remains in me.

1.2.2.3 *Calligraphy handwriting couplets and my first taste of knowledge*

My parents were the most knowledgeable couple in the village (Fig. 1.16). My father, a primary school teacher, was the most knowledgeable man in the village. My literate progressive mother was the most educated woman (a graduate of high primary school, like middle school nowadays) in the

Fig. 1.16   My parents at the door of our house with parts of couplets handwritten by my father in 1983.

village. My siblings and I were fortunate to have our first taste of knowledge through them. My father habitually wrote the calligraphy couplets in large characters at the entrance of our cave house and on the door of each room in the house. He was also asked to do this for many of the families in the village every year before the Chinese New Year.

The calligraphy couplets, written boldly by my father using the traditional Chinese brush pen and posted on the central wall at home, described a shining golden house filled with wide knowledge and endless ocean with clear water through an ancient Chinese proverb, *"Only through hard work and diligent labor one can acquire a golden knowledge house and a ship that could travel on the wide ocean"* (书山有路勤为径，学海无涯苦作舟). As a child, I found the couplets fascinating and attractive, as they reminded me of the importance of studying hard and helped me become a hardworking person.

I studied as hard as I could, but my behavior gave my classmates some jokes. For instance, I said "sorry" to a tree during my reading walk, I ate food with the writing ink on it. It was really difficult to study well when one was near starvation. However hard my parents worked, there was not enough food or clothing. Nevertheless, my parents did their best for their children, so that the children could remain alive even in the hardest of the times.

1.2.2.4 *Making little money from the wild ground and trash*

Due to the poor economic situation, typically no money could be given to the children, except for the lucky money (Chinese red envelope gift), less than 1 Yuan per gift, given at Lunar New Year's Eve by my parents, grandfather and close relatives. The giving of lucky money to children during the Chinese New Year is a transition that is still practiced today.

As farmers usually did in the autumn, we obtained dandelion and capillary artemisia by digging the wild ground. Through the digging, we also collected the cypress seed, wild jujube wolfberry and some traditional Chinese herbal medicine. Then we used these items to exchange for some money in the Chinese medical shop. In addition, we picked up the old flexible aluminum toothpaste tube container, rubber shoe soles, and recycle waste from the trash, to be sold in the waste collection station. Thus, I was able to earn some coins, which enabled me to buy a Chinese dictionary. I had wanted to have a dictionary for a long time during my years in the middle school. This was the first dictionary among my classmates. My teacher praised me for this. Moreover, the dictionary enabled me to learn many words by myself without having to ask my teachers.

This experience with the wild ground also induced some questions in my mind. Why can the seemingly useless plants, even grass, serve as medicine for healing the sick? Why can edible oil be made from trees, particularly the sesame tree? However, I found no answer, not from the interesting comic books or any of the books that I could find.

1.2.2.5 *Guessing at the next lesson*

In class, I was often able to imagine, guess and recognize correctly the next lesson. Therefore, I was effectively reviewing during class, as opposed to previewing the lessons before class. In this way, I was able to read more carefully and imagine the various possibilities more freely, in addition to checking the results in the next class. In most cases, I obtained the results correctly. This gave me a feeling of satisfaction and the consequent happiness. Moreover, this method of study aroused my enthusiasm for study and made me feel a bit like of a scientist. This helped me start to dream of becoming a scientist in search of interesting scientific principles.

I respected my teachers greatly. While I was in middle school, I was effectively attached magnetically to many of the courses. The phrase "*like a bee to a flower, a bread to the hungry*" was used by some of my classmates to describe me. My classmates even wrote the phrase in their homework, which was then displayed in the classroom by my teachers.

1.2.2.6 *My scientist dream and the historian of 2000 years ago*

In the first grade of primary school when I was 8 years old, I learned in class some words corresponding to various occupations, such as engineer, farmer, soldier and scientist. I remember climbing an apricot tree and staying at a tree fork, while I discussed with my classmates under the tree what we wanted to be when we grew up. I immediately chose scientist for my future occupation and proudly announced my scientist dream.

As a pitiful little girl, my unusual dream to be a scientist was laughed at by my childhood friends due to the lowly status of women. However, I was encouraged by a sentence from a Chinese text that was read by my sister who was only two-year older than me, "*How could a bird know my fierce ambition to become a large swan?*" (燕雀安知鸿鹄之志). This text was from the famous book (史记 - 陈涉世家) "Historical Records — Chen She's family", by Sima Qian (司马迁, Chinese historian, 145-90 BC, the early Han dynasty, more than 2000 years ago) (Fig. 1.17). The significance of Sima Qian to the Chinese people is akin to that of Herodotus to the Europeans. It was surprising that Sima Qian became my soul mate, as he seemed to be able to understand me in spite of his being outside my world in both time and space. I looked forward to knowing more about him by reading and studying hard in the school. I also recognized the importance of knowledge.

Most of girls in the village stopped their school life just after their primary school, but I never gave up, even when life was very hard, with inadequate food, clothing or medicine. I simply tried to do my best.

Fig. 1.17   Sima Qian, Chinese historian (https://en.wikipedia.org/wiki/Sima_Qian, public domain)

1.2.2.7 *Making exercise books from the waste packaging*
paper obtained from a cement factory

Due to the shortage of money, my family could not afford to buy exercise notebooks for my use in writing. However, I did not want to give up practicing writing when I was in primary or middle school. Thus, I made exercise books myself by collecting the waste kraft packaging paper from a nearby cement factory. After my manual cleaning, cutting and tailoring, I used the paper for writing and even for homework. Fortunately, my teachers were kind and did not criticize me for using such paper. In particularly hard times, in order to save the paper, I wrote on the waste tree branches on the soft loess ground. The writing luckily disappeared due to the usual wind or rain.

In China after 1977, it was possible though examination for people to study in a university. This was when I graduated from primary school. The competition for entering a university was so keen that only a few students, typically the top 1 to 3 in a class, could pass the difficult entrance examination. Although I was in the top 3 of 400 students in 8 classes in middle school and in the top 10 in 8 classes in my grade in high school, I felt disillusioned and avoided thinking of whether I could pass the university entrance examination or not.

1.2.2.8 *Longing to travel*

Travelling was difficult and costly in my childhood. As one of four children and the second girl in the family, I rarely went outside the village, whether the travel was by walking or other means of transportation. Thus, I had never visited museums or other places outside my village, though I had attended some weddings nearby the village by walking there. Particularly memorable was my aunt's wedding, which involved a carriage for the bride. Nevertheless, there were some country fairs in a small town that was 2 km from my village. Through these well-attended fairs, I became aware of the life of people outside my village. My knowledge mainly came from my teachers and parents, in addition to movies and books. The longest river in China, Huanghe, is only 200 km from my village, but I did not have any opportunity to see it throughout my childhood. I was attracted to travel stories and longed to travel and see the wide world. Striving to enter a university, I understood that my only path to achieving my scientist dream was to do well at the university entrance examination. However, due to my underpriviledged situation, I felt disillusioned. In spite of my dream, I did not think about entering a university, nor did I think about becoming a university professor.

I was drawn to the concept learned in my biology classes about the possibility of changing the gene. Thus, I imagined changing my gene through studying hard, so that I would become much cleverer. I thought that cleverness would help me in whatever I wanted to do, including farming, and might be inherited by my next generation.

Due to the strict discipline of my father, a primary school teacher and the most knowledgeable man in the village, and the loving care of my literate progressive mother, the most educated woman in the village (with education through the high primary school, i.e., the middle school now), I was able to attend and do well in primary, middle and high schools. This laid the foundation for my later university education, which was preceded by a challenging university entrance examination. In the 1980s, fewer than 10% of the students that took the examination were admitted to a university.

1.2.3 *Studying in the university*

1.2.3.1 *Moving from the village to the university*

In September 1983, I entered Changchun College of Geology (presently College of Earth Sciences, Jilin University 吉林大学), a key university of China located in Changchun (Fig. 1.18), that is more than 2000 km from my village. I majored in Geology because I like the wild nature and hope to understand and explore the new knowledge in it.

Before I left for the university, my parents gave me the best clothing in the family, including a shirt worn by my elder sister at her wedding, trousers handmade by my mother, the first commercial waist belt made of canvas bought by my parents in the shop, an old suitcase borrowed from my maternal aunts, to pack my belongs, and about 30 cakes as the meals of my trip handmade by my mother one day before my leaving. I shall always remember the love and support from my family.

In the university, I was excited by the new world in front of me, particularly the knowledgeable professors, and the classmates from different provinces and even different countries. The huge library impressed me. I was able to read books without having to pay anything. Thus, the realization

(a)   (b)

Fig. 1.18 Attending university in Changchun College of Geology (presently College of Earth Sciences, Jilin University). (a) Visiting the Tiananmen Square in Beijing in 1983. (b) Reading in my dormitory in Changchun College of Geology in 1986.

of my scientist dream became a possibility. I studied hard and enjoyed the work very much.

1.2.3.2 *Being inspired by some successful people*

As I grew up, I was inspired by the lives of a few people who succeeded in their university education. The inspiration helped me, since I had no idea of how to realize my scientist dream.

One day I heard for the first time the title of doctor. It was from a radio broadcast concerning Dr. Henry Kissinger (1923-, U.S. Secretary of State in 1973–1977) (Fig. 1.19). He became very well respected and famous later in China. I did not know him at all, and I did not really care about who he was and what he was doing. However, I admired somebody with the title doctor and the respect associated with the title. Thus, since I was a child, the title of doctor was in the back of my mind. I did not understand the connection between university and doctor, but the seed of doctor as a knowledgeable person was planted in my young mind. Of course, we knew more about Dr. Kissinger later as the first person that came from the U.S. to China after the War of Liberation. He came to China in the 1970s and received the Nobel Peace Prize in 1973.

The second successful person that inspired me was Ms. Yi Wu (吴仪, 1938-) (Fig. 1.20). A newspaper with her photo and intro-duction was on a cupboard in my high school. She graduated from

Fig. 1.19   Kissinger (left) with Enlai Zhou 周恩来 (center, 1898–1976, the first Premier of the People's Republic of China) and Zedong Mao 毛泽东 (right, 1898–1976, the founder of the People's Republic of China), negotiating about rapprochement with China.
(https://en.wikipedia.org/wiki/Henry_Kissinger, public domain)

Fig. 1.20 Ms. Yi Wu (left) with U.S. Secretary of State Colin Powell (1937-).
(https://en.wikipedia.org/wiki/Wu_Yi_(politician), public domain)

China University of Petroleum with a degree in Petroleum Engineering. She later became the Vice Premier of China (2003–2008), being best known for her work as Minister of Health in 2003 during the SARS outbreak.

I was also inspired by Professor Jingyin Li (李景荫) (Fig. 1.21), who studied in Furen University (輔仁大學, a part of Beijing Normal University 北京师范大学) in Beijing in the 1940s, before the War of Liberation. He was from a nearby village, with parents who could not afford to send him to a university. Nevertheless, he passed the university entrance examination of Furen university. To help him with his university living expenditure, various families in his village donated silver dollars. He studied with great success and eventually became a top professor in Renmin University of China (中国人民大学). My primary school teachers showed us his beautiful calligraphy made using a thin brush pen when he was in primary or high school.

My father and I visited Professor Jingyin Li in Beijing on our way to my enrollment in Changchun College of Geology in the autumn of 1983. After some free talk with my father, Professor Li advised me, saying, *"Being an undergraduate student in a university is like being a primary school student. If one wants to become an expert in a field, one has to go on to study as a graduate student so as to obtain the master's degree and doctorate degree."* Thus, even as I embarked on my undergraduate study, I had in mind study toward the doctorate degree. I greatly appreciated the advice from Professor Li (Fig. 1.21), in addition to the support from my father at the beginning

Fig. 1.21   Prof. Jingyin Li 李景荫 (center), my husband (Jianhai Li 李建海) (left) and me (right) in 2008.

of my university life, and also later when I became a teacher in Peking University.

Another person that served as my role model is Prof. Hongwen Ma 马鸿文, who succeeded to enroll in Chengdu College of Geology with the recommendation from the people of his villager without taking the university entrance examination in the 1970s and became a famous professor later. I admired him for his opportunities to travel to many places for work after he enrolled in China University of Geosciences in Beijing. In particular, he visited the wide fields in Xinjiang, which seems to be rich in sweet grape. I had wanted to visit there, partly because one of my elder cousins (a daughter of my uncle) worked there and I had never been there.

1.2.3.3 *University life*

As a university student, I was supported financially as much as possible by my family. However, life was not easy for an 18-year-old rural girl from a poor village. Compared to my university classmates, nothing was competitive for me, except for my academic record and running skill. Yet, there were sweet memories of my university life.

1.2.3.4 *My clothing*

I wore almost all the time during the four years of my undergraduate education the shirt that my elder sister gave me (Fig. 1.22(a)). The shirt was worn by her at her wedding and was the most favorite clothing in her life. Over the four years, the shirt became so worn out that the regular rectangular corners at the front of the shirt decayed and became folded, due to the repeated mending, which reduced the edges gradually to irregular triangles. This shirt became my embarrassment, my symbol, and a joke among the boys in my class.

Some of my classmates shared with me their belongings, such as sunglasses, which I wore for photos at times (Fig. 1.22(a)). I often wore the T-shirt for the university sports team, even though the shirt was meant to be worn during sports training only. This shirt was my best clothing (Fig. 1.22(b)).

For a rural person, waste belts are typically made by hand using old worn-out clothes or leftover cloth. My first canvas belt was bought for me by my parents before I departed for the university. I used it so much that it was worn out and I had to mend it repeatedly. I could not afford to buy another one. There were several blocks between the dormitory and the classrooms and laboratory. I often worried about the possible breaking of the belt during my rush running or slight sneezing.

(a)   (b)

Fig. 1.22 My clothing in university (a) Wearing the shirt given to me by my elder sister, who wore it in her wedding. (b) Wearing my sports T-shirt (front left 1).

I still remember clearly viewing the belt display counter at a shop entrance, wishing that I could afford to buy one. This memory keeps me mindful of not wasting any belt, even today.

In spite of my limited clothing, I felt fortunate compared to the poor peasant girls in rags. These girls were unable to go to school. Instead, they performed hard housework and got married at a young age.

1.2.3.5 *My food*

Because of my poverty, only the cheapest spicy cabbage, tofu and kelp could be afforded in the university canteen every day. Such food was mixed with the fried spicy chili powder (Fig. 1.23(a)), like today's popular commercia Lao Gan Ma or Old Godmother chili sauce (Fig. 1.23(b)), but with only a small amount oil, that my parents prepared for me at the village. These constituted my standard meals, during my four years in the university. My classmates considered these as my "favorite" and my food style. I was hospitably entertained and hosted with this kind of food by friends for many years after I graduated from the university.

1.2.3.6 *Following the advice from my parents*

My parents unceasingly advised me, saying, "Look forward to the top level in studying, in spite of the low level in life". I just tried to do my best, studying very hard, enjoying the reading in the library, and exercising in the sports playground, without comparing my life (in terms of clothes, food, tour, etc.) with the typical life of others around me.

(a)   (b)

Fig. 1.23   (a) Fried spicy chili powder. (b) Commercial fried chili sauce.

1.2.3.7 *Buying an English-Chinese dictionary – my luxury*

After visiting a bookstore counter numerous times, I bought an Oxford English-Chinese Dictionary by using almost a month of my living expense budget. To me, this was the greatest luxury of my life so far. For this purpose, I decreased my food expenditure for at least half a year.

There were some big public dictionaries in the library, but they were far from the classrooms and dormitory. As a consequence, I had to collect the new words and then check the meaning of the words in the dictionary in the library about a week later. The new dictionary was indeed useful, as it facilitated my reading of reference articles in English and helped me improve my English. I kept this dictionary in my book bag every day, thus making the bag heavy and causing me the nickname "Doctor", even though I was just an undergraduate student. This nickname reinforced my dream of obtaining a doctorate degree.

1.2.3.8 *My weekends*

In the weekends, I usually ran in the sports ground and did physical exercises for the sake of my health. In addition, I relaxed by walking around and enjoying the beautiful bright golden flowers that bloomed on the wild hills near the campus. I also enjoyed the plants and beautiful flowers in the public garden, and I spent some time in the big stores and small markets in the city, in spite of the little money that I had. These activities helped improve my health, open my eyes, extend my knowledge, and understand more about the city and the world.

1.2.3.9 *Like a hungry bee in the library*

My favorite place for spending my spare time was the library of the university. There I could read many books freely like a hungry bee around honey, without any payment. The books included specialized books in my major discipline of study, as well as novels and modern and ancient books in Chinese and English. Reading was enjoyable and helpful to my study and future career. I have read almost all the specialized books in the library. Along with the reading, I took notes. I read not only the books recommended by the professors in the classes related to my major discipline of study, but also books in other disciplines, such as those on natural and social sciences, and the biography of outstanding scientists, such as Marie Curie and Thomas Edison, whom I had long respected. The reading helped me make early progress in my studies and provided me guidance in life, including marriage.

My daily schedule was mostly occupied by the numerous courses that I took. My schedule was relatively open in the evenings and weekends. However, I wanted to use the library as much as possible. After my numerous requests, a tall librarian agreed that I could stay in a locked reading room in the library. Thus, I was able to spend long times, even during holidays and vacations, to read books in the library, like a bee sucking in the nectar from flowers.

1.2.3.10 *As a running champion having caught the attention of a motherly advisor*

Due to poverty, rural children typically had little choice of hobbies. Instead of dancing, singing and music, long distance running was my hobby. I was able to run fast because of the long distance between my home and my school in my village. The distance forced me to walk or run fast throughout my childhood. It was quite difficult for a kid, even a teenager, but it was good practice for a runner. Due to my ability to run, I was asked to enter the sports school in lieu of the regular high school. However, I did not want to give up my scientist dream.

Due to my high running speed, I took part in the sports game in the city as the representative of my university. In addition, I won the running champion prize in my university for numerous times (Fig. 1.24), and also held the speed record of the university.

My mother often said, *"No effort is useless"*. This is indeed true in relation to my running.

It was a big surprise for me to be looked for by a teacher, M. Zhang, after the announcement that I was the champion of a long-distance running race in the sports game for new students. As my enrollment officer, Miss Zhang found my name and my stated hobby of long-distance running caught her attention. She then invited me to enjoy her steamed stuffed buns with her in some holidays, and sent me small heart-warming souvenirs, such as a red apple, when I left. In addition, like a mother, she kindly gave me advice

Fig. 1.24   Winning a long-distance running race in the university.

about clothing, cooking and so on. The most important thing that she did for me was instilling in me self-confidence, which was in contrast to an inferiority complex that might be typical for a lonely rural girl.

1.2.4 *Financial challenges*

1.2.4.1 *Financial burden on my family due to my university education*

As a primary school teacher in the village, my father earned very little money. It was difficult for my parents to support my university education, in addition to the school education of three other children.

The ticket for the trip to the university was very expensive for my family. In order to raise the needed fund, my parents sold some crops and eggs, and asked for donation from my aunts and grandfathers in the amount of less than one hundred Yuan each. Furthermore, my mother walked about 10 km in order to borrow money from some relatives. However, the amount raised

Fig. 1.25   The deep-fried dough sticks (Youtiao 油条) were prepared by my parents and
sold in the market with the help of my sisters and younger brother.

was not enough, so my father took a loan from a bank to pay for my lodging in the university. My admission notice letter from the university was used as a basis for obtaining the loan. Finally, I departed for the university with many cakes that my mother cooked a day before the trip. Soon after I had graduated and obtained a job, the loan was paid off.

In order to pay for my lodging and living expenses, my parents had to do many odd jobs, such as preparing the deep-fried dough sticks (油条) (Fig. 1.25) and selling them on the ground in the market with the help of my sisters and brother as waiters.

Very little money was left for the family's own use, as most of the income was kept in a jar for paying for my living expenses in the university every month. It was really a heavy burden for a rural family to support a university student.

While I studied in the university, I did not have enough money for food or other necessities. Knowing my difficulty, a university student in a neighboring village advised my father to provide me with more money. However, I never complained to my parents, because I understood their financial situation.

As Professor Li (Fig. 1.21) mentioned, he had a very difficult time when he was a university student in the 1940s, in spite of the donation from his entire village. He further advised me to remember the villagers who had helped me and give them at least double of what they had provided.

1.2.4.2 *Receiving the scholarship for the top university student*

I tried to find a part-time job to support myself while I attended Changchun College of Geology. However, there was no chance for a girl like me from the rural country. My job application was turned down even by the construction sites.

Without any hope for a part-time job, I concentrated on my study, hoping to receive the scholarship for the top student in the university. I thus won every year the scholarship, which provided 50 Yuan as the total amount for a year. In addition, due to my poverty, I received the maximum amount of 21 Yuan per month as subsidy for my living expenditure. The scholarship was greatly beneficial to me as a student. It also helped my career after I had graduated.

1.2.4.3 *Being spared from hunger as a result of being in the university sports team*

Due to my fast running, I was selected to be an athlete (a runner) in the sports game. Thus, I received training as an athletic member of the university sports team (Fig. 1.26(a)) and of the female team (4 × 400m) (Fig. 1.26(b)). This enabled me to receive additional subsidy, as the food ticket amounts were calculated according to the time spent on the training.

Most of my teammates attended the training just for improving their skills in the sports. In contrast, I did so mostly for the food tickets needed to sustain my life. This intention was my secret throughout my years in the university. I considered the secret to be shameful, but the food tickets spared me from hunger. Moreover, the sports training was good for my physical health.

1.2.4.4 *Social pressure in the village against the education of girls*

It was customary thinking in my village that one should not spend money to educate girls, as they would get married and become members of other families. Numerous relatives clearly expressed this to my parents. They had even said this to me directly. In response, my father said to me, "*I promise to support all of the education that you can reach*". What a great father!

In order to change the old customs that were disadvantageous to girls, I strived to be a role model for girls through performing excellent work. I also tried to influence parents who did not send their young daughters to school. My hope is that more rural girls would go to school and receive good education, so that they can contribute to society, like the boys, at least to be the best of their abilities.

(a)

(b)

Fig. 1.26   In the university sports team (a) (front right 3) and the female team (4 × 400m) (b) (right 1).

1.2.4.5 *Advancing in my scientific career*

With a grateful heart for my wonderful parents, I studied very hard all the time in the university. Thus, I received my bachelor's degree in Geology in 1987 and my master's degree in 1990. My master thesis is on "The

structure of natural porous rock, pumice in Changbai Mountain 长白山".
Both degrees are from Changchun College of Geology 长春地质学院 (fore-
runner of College of Earth Sciences, Jilin University 吉林大学). In 1990–
1994, I was a Lecturer teaching Mineralogy in Xi'an Institute of Metallurgy
and Architecture (presently Xi'an University of Architecture and Technology 西安建筑科技大学) in Xi'an, Shaanxi. In 1994, I entered China University of Geosciences (Beijing) (中国地质大学), where I received the Ph.D.
degree in 1997, with a thesis on "The synthesis, structure and transport
properties of The $CuCl_2$-$NiCl_2$ intercalation compounds of the expanded
graphite". After that, I worked as a Postdoctor with Professor Hejun Li
李贺军 as the advisor in Northwesten Polytechnical University 西北工业大学
in Xi'an. In 1999, I joined the School of Earth and Space Science of Peking
University in Beijing as a Postdoctoral Fellow with Professor Zhe Zheng
郑辙 as the advisor, researching in the microstructure of natural mineral
materials. In Peking University, I was promoted to be Associate Professor
in 2000 and Professor in 2005, with research focused on mineralogy and
mineral materials science.

It was a long journal of study and research. However, I am gratified to
be given the opportunities, which led to my receiving a number of awards,
including the Best Youth Scientist in Beijing in 2001 and 2005, Houdefeng
Prize (侯德封奖) for the Best Youth Scientist of Chinese Society of Mineralogy and Geochemistry in 2004, the Zhengda Prize (正大奖, for the excellent
teachers in Peking University), the Longruan Technical Prize (龙软奖, for
the excellent teachers in the School of Earth and Space Science of Peking
University), the Best Teacher of Peking University in 2005 and 2015, the
Best Ph.D. Thesis Advisor of Peking University in 2016, the First Prize for
fundamental research in Nonmetallic Mineral Science and Technology in
China in 2021, and so on. I have always considered teaching to be a very
important component of my work.

1.2.4.6 *My siblings' education*

To pay for my lodging in the university, my mother did more farming work
in order to get more harvest, and my father worked much harder in his
teaching. They also did numerous odd jobs in 1980s with the help of my
young brother, my elder sister. All of family members took part in the
farming work (Fig. 1.27(a)); even a little boy (Rui Chuan, 传睿), the son of

(a)

(b)

Fig. 1.27 The family photos were taken during my university. (a) In support my lodging in the university, my mother Zhulan Wang 王竹兰 (left 1), and father Chun Chuan 传春 (center) made great effort to farm, with the help of my young brother Jianwu Chuan 传建武 (right 1), my elder sister Yayun Chuan 传亚云 (left 2), and her little son, Rui Chuan 传睿 in 1985. (b) The family photo taken in the Chinese New Year of 1984, with my mother (front left), my father (front right), my younger brother (back center), my elder sister (back right), and my younger sister, Lingyun Chuan 传凌云 (front center).

my elder sister, also enjoyed peeling the corn leaves and together with the family in front of our cave dwelling in 1985.

Due to the poor economic situation, it was very difficult for my parents to afford more children to the university. I was the only child that was given the opportunity to study in a university. All my siblings had to give up their university dream in disappointment. But the cherished heritage, particularly cultivation (quality improvement) and reading, still worked in the mind of everyone in family. All of us were happy to share the smell of university as we took a photo in the Chinese New Year of 1984 — my first family reunion after entering the university in 1983 (Fig. 1.27(b)).

My father received his bachelor's degree in Dali Normal College 大荔师范学院 by self-study after I had graduated from my university in 1987. All my siblings worked hard to earn money for their education. My elder sister (Yayun Chuan 传亚云) gave up taking the university entrance examination, even though she was among the top 10 in her class and some of her classmates succeeded in entering universities. Instead of studying in a university, she attended a professional high school 澄城职业技术中学 to study the electrical technology, received training in computer programming in Northwestern Polytechnical University 西北工业大学 in Xi'an, and Changcheng Computer School 长城计算机学校 in Beijing, and then became a successful factory worker. After retirement, she attended a university for the elderly only for knowledge enhancement, though she was still trying to make up for her lost dream.

My younger sister (Lingyun Chuan 传凌云) hid her university dream. Although she was the top student in her middle school, she did not attend a typical college, but attended instead a typical technical school, Shaanxi coal industry health school 陕西煤炭工业卫生学校. Nevertheless, she spent her spare time after her salary work to study in Xi'an Medical University 西安医科大学 and Northwest University (西北大学) by herself only for knowledge enhancement, thereby receiving her bachelor's degree. Later, she became a famous writer and served as the Vice Chair of Weinan Writers' Association, the Vice Chair of China Literature and Art Critics Association (Weinan) now, the former vice chair of China Federation of Literary and Art Circles (Weinan).

My younger brother (Jianwu Chuan 传建武) also attended a professional high school 冯原职业技术中学 for tobacco planting. He received professional vehicle driving and computer training in electronic commerce for the agriculture products and later became an excellent progressive farmer. With some professional training, he also became a salesperson of health insurance and for China Unicom (a large telecommunication company).

All of the grandchildren of my parents received university education in the recent years. Some of them even received their Ph.D. degrees. The youngest granddaughter of my parents is studying in a key university and pursuing her dream now.

It is gratifying that my family members strived in their education, in spite of their difficulties. I am proud of them. The smell of cultivation and reading still persisted in our minds, whether in a city or the small village, whether life was difficult or not.

1.2.4.7 *My gratification*

I would not have been able to complete my university education without the sacrificial support of my family and the scholarships from the university for good students from poor families. My parents deserve my utmost thank, because they sacrificed for me amidst starvation. I also greatly appreciate my sisters and younger brother, who allowed me to use the most needed materials. My warm regards also go to the folks in my small village, as they helped me and even saved my life in the times of danger, particularly that old man who gave me a hand at the deep well edge while I was enjoying the water surface of the deep well as an interesting mirror in the eyes of a little naughty girl. I shall continue to make every effort to improve the situation in the village, where the life is quite difficult even now.

1.2.5 *From friendship to marriage*

Since I was a child, my classmates and friends, whether boys or girls, could visit me without much limitation from my parents. We could play together freely and discuss and find solutions for hard questions. When we could not persuade each other after hot arguments, even after serious quarrels, we sought help from our teachers. However, we communicated well, made good progress together, and became friends.

Most of my childhood friends remained my friends even after many years. One of them (Jianhai Li 李建海), became my husband in 1990 (Fig. 1.28). After he graduated from Northwest Agricultural University in Xi'an, he became the agronomist in a farm in a suburb in the western part of Beijing. We formed a happy family, with good communication and

Fig. 1.28 One of my high school classmates and I were married in 1990.

cherished times of relaxation. We have one child, a son, who was born when I was 27 years old.

Marriage is wonderful, as the husband and wife can support one another in both good times and bad times. One should not be shy to make friends, whether the friends are girls or boys. Especially in the countryside of China, girls are commonly supposed to have girlfriends only, thus resulting in lopsided knowledge that is not beneficial for the later life. The world is wide enough to accept girls and boys studying together and men and women working together under the same sky. Since I was a child, I was brave to communicate and make friends with boys. Knowledge is not compartmentalized for boys and girls. Girls also are human beings who can share the world with boys. Together boys and girls can improve the world.

1.2.6 Scientific research

1.2.6.1 The essence of scientific research

Research is the key for a scientist to make an impact to the world. Success in scientific research requires strong grounding in science. The grounding involves firm understanding of the basic scientific concepts and broad knowledge of subjects that are not limited to a particular discipline. No foundational knowledge is adequate for sustaining one's research for more than a few years. Therefore, one needs to study hard not only when one is a student, but one must study continuously throughout one's career. Creativity is also necessary, particularly since the publishing of research results requires the results to be new in the sense that such results have never been published by anyone in the world. Thus, one must be aware of the status of science, so that one knows what has been published and what has not been published. In addition, one must be knowledgeable about the needs of technology, so that the science is pursued with relevance to applications.

1.2.6.2 Beginning my scientific research career

Due to my hard work and research publications, I secured my dream teaching position as an Associate Professor in Peking University. However, this was the beginning of my facing a higher level of challenge. I understood that this position would provide a route for my reaching my dream to become a great scientist like Curie, Edison, Mendeleev and so on. In spite of my enthusiasm, I faced many difficulties, due to inadequate research funding and facilities. For a couple of years, I did not have office or laboratory space. Getting research funding was very difficult. The situation made it difficult for me to attract graduate students. Therefore, I was very anxious about my future. The difficulties continued for several years.

It is never easy to start a new scientific research operation. For me, it was extra difficult. Due to the absence of laboratory space, I used spare classrooms as temporary laboratory space. Thus, I had to move my laboratory set-up frequently. I sometimes shared space and equipment in the public student laboratory with colleagues when the students were not using the laboratory. About two years later, I was provided a small room as my laboratory, Then I started to build my laboratory with used equipment, containers and tables. Using such primitive facilities, I performed some experiments, and compared the results with those of researchers in other parts of the world.

1.2.6.3 *International research cooperation*

The attending of international research conferences is valuable for widening my knowledge and providing me with opportunities to know and discuss with researchers from various countries. However, the conference registration fees and travel costs were too high for the limited funding that I had. I thus contacted some of the conference organizers and was able to obtain a discount for the conference registration fee. Although I paid for my travel, I sometimes obtained meal tickets along with the waiving of my registration fee. I am grateful for the organizers' understanding and help.

The interaction with the international researchers was valuable not only for my research, but also for my teaching. It was wonderful to get to know some of the scientists from various laboratories in the world. The relationship enabled me to engage in international cooperative research with prominent scientists, in addition to organizing some official international cooperative research projects. An example is Madame Agnes Oberlin (1910–2019) of CNRS (The French National Centre for Scientific Research, which is among the world's leading research institutions in Orleans, France) (Fig. 1.29(a)).

Madame Oberlin was an international leader in analyzing the structure of carbon materials by Transmission Electron Microscopy (TEM). She is featured in Chapter 10 of Vol. 1 of this book series. Our international relationship helped me obtain joint financial support from French and Chinese research funding agencies to perform the cooperative research. In addition, I received valuable TEM training from Madame Oberlin (Fig. 1.29(a)), and worked with her group in Orleans, France (Fig. 1.29(b)). She also visited my laboratory in Beijing and advised me on my research.

Professor Jean-Baptiste Donnet (1923–2014) (University of Haute Alsace Scholl, France), a pioneer in the surface chemistry of carbon black,

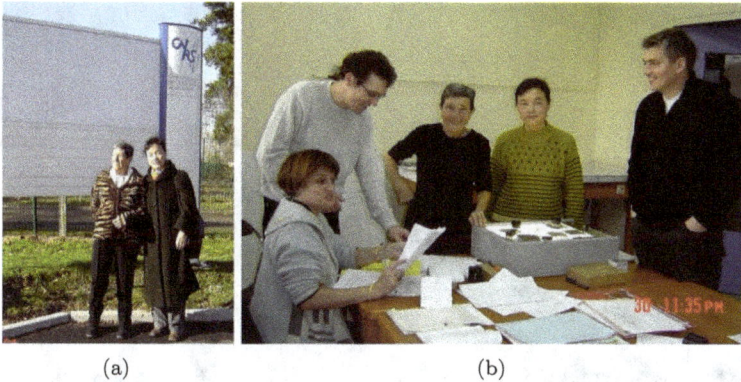

(a)   (b)

Fig. 1.29 Working in an international cooperative research project with Madame Agnes Oberlin (right) in 2004 (a), and discussing with her group, Dr. Sylvie Bonnamy (front left 1), Alain Pineau (back left 1), Madame Agnes Oberlin (center), Thomas Cacciaguerra (right 1) and me (right 2) (b).

was invited by Madame Oberlin to engage in the research program as the host of a joint project in France. With an experiment involving the lighting of a candle in his office, Professor Donnet explained the formation of carbon black very clearly (Fig. 1.30(a)), thereby enabling us to explain our experiment results well (Fig. 1.30(b)).

With the international grant, I was invited to give presentations in Jeonju University (Fig. 1.31(a)), Chungnam National University and Chonnam National University in South Korea. In addition, I worked in Jeonju University as a visiting professor in 2016. It was nice to join their forum with the scientists in Jeonju University (Fig. 1.31(b)). The communication with the scientists, such as the top Korean scientist, Professor Seung Kon Ryu, was interesting and helpful. The discussion with businessmen was enjoyable. Our discussions covered not only research, but also life (Fig. 1.31(c)).

In addition, through joint funding from German and Chinese research funding agencies, I was able to conduct research in the laboratory of Professor Peter Scharff (the rector of Universität Ilmenau in Thüringen, Germany). Furthermore, a top Japanese scientist, Professor Michio Inagaki, invited me to his university in Japan to work for a year as a visiting researcher. A top Brazil scientist, Prof. Francisco G. Emmerich, also invited me to his university, Federal University of Espirito Santo in Brazil.

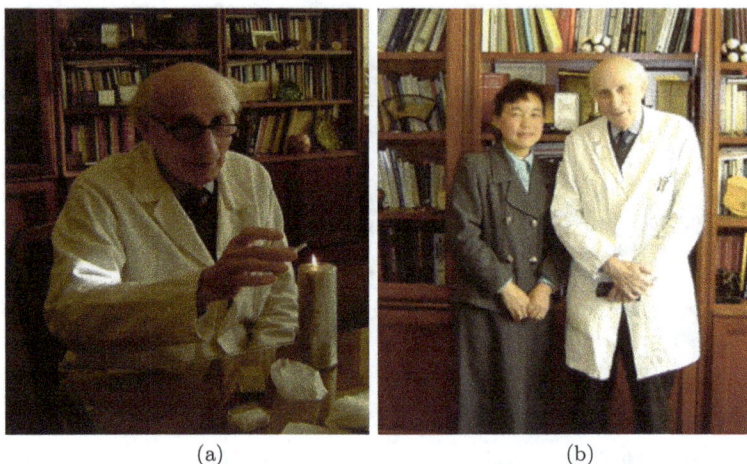

(a)   (b)

Fig. 1.30   Professor Jean-Baptiste Donnet (1923–2014) (University of Haute Alsace Scholl, France, a pioneer in the surface chemistry of carbon black), explained the carbon black by lighting a candle in his office (a) and gave help to explain our experimental results (b) in 2004.

These international interactions widened my horizon and enabled my research to be connected with that of the world. More recently, my interaction has been extended to researchers in Russia, Brazil, Czech, Austria, Spain, the U.S. and so on.

1.2.6.4 *Study and knowledge*

"*Why study?*" is an important question (Fig. 1.32). It seems to be the motto of University of Oxford. "*Why study? The more I study, the more I know. The more I know, the more I forget. The more I forget, the less I know. So why study?*".

We often try to find a good answer to encourage ourselves and to give the answer to children.

"*Study is for the bad memory*" for the elderly, Prof. Kap Seung Yang answered.

"*Study is for the higher life skill and better life*", the answer also is clear in my mind. It looks like the proverb "*knowledge changed the destiny*" (知识改变命运) in Chinese.

"*Why study, why research?*"

Fig. 1.31   Working as a Visiting Professor in South Korea. (a) Giving a presentation in Jeonju University, (b) Joining the forum organized by professor Hong-gun Kim (front right 5), (c) Discussing with the group of Professor Seung Kon Ryu (right 2) Prof. Young-seak Lee (right 1), and Mr. Kil Wan Lee (CEO of a Korean company (left 2)), (d) enjoying the work with the group of Prof. Kap Seung Yang (right 2) after the presentation in Chonnam National University.

Fig. 1.32   The enjoyable important question, *"Why study?"*

"*It is for the better life*", not only for oneself, also for human beings, also all the lives. The answer also is clear in my mind.

1.2.6.5 *Steering my scientific research*

The selection of a research topic is critical to research success. With tears in my mind, I pondered over questions such as "*What is the topic? What I could and should do?*", for quite a few years. What I learned in the courses that I took only gave me the foundational concepts that are associated with the knowledge that existed at the time. These concepts are important, but they do not usually point to particular scientific questions that are open and need to be addressed. The scientific questions need to relate to fundamental science, not just applied science. In addition to recognizing the open scientific questions, one must connect the science to the needs of society and the industry. For example, industrial needs include decreasing the cost, energy requirement and environmental impact of product fabrication, increasing the performance and durability of the products, and improving the feasibility of recycling the materials after use. Galileo Galilei (1564–1642) said, "The only purpose of science is to alleviate the suffering of human existence. Scientists should think for the majority of people".

The research must be relevant to both science and technology in order to the research to be fundable. Without research funding, it is very difficult to perform research, particularly when the research involves experiments. Thomas Edison (1847–1931, American inventor) often took the initiative to raise the funds for his experiments. With the money from the patents, he performed new experiments and greatly expanded the use of electricity. Madame Marie Curie (Fig. 1.33, 1867–1934, Polish naturalized-French physicist and chemist) was often short of funding too and needed to beg to raise the fund. One must not be shy in asking for financial assistance.

In addition, with a research goal in mind, one must devise the research approach, which must be sound and feasible, in addition to being effective for reaching the goal. The competition for research funding from the National Natural Science Foundation of China is very keen. Eventually, I decided to choose the topic according to what interested me and what I was capable of doing. My capabilities relate to my skills as well as the facilities available to my use. As I pursued the research, I continuously reminded myself to persevere and not to give up my dream.

Fig. 1.33   Madame Curie (https://en.wikipedia.org/wiki/Marie_ Curie, public domain)

A scientist may choose to be a pioneer in a certain field and let other researchers from various parts of the world follow the research direction that has been opened. Alternatively, a scientist may be a part of a large research group that is led by one or more successful scientists. In the latter route, the scientist is a member of a research group and the topic selection is made by the leader of the group, thus the risk of the research is relatively low. The former route in which one is the pioneer is more challenging and much riskier, and requires more courage, creativity and leadership ability. However, if the pioneer is successful (usually after numerous years of research), the impact of the work tends to be greater. I try to operate with a balance between these two routes.

In operating my research program, I supervise the graduate students in my research group by choosing their thesis research topics and advising them as they progress. Using the research results to support a meaningful research direction, I applied for research funding from National Natural Science Foundation of China. It was difficult and I experienced failure in securing the funding. However, after strengthening the research results and improving the research proposal, I eventually succeeded in securing the funding.

1.2.6.6 *Receiving help from the experts*

I greatly benefited from the understanding and help from some prominent scientists. Madame Agnes Oberlin (Fig. 1.34) visited my laboratory for a

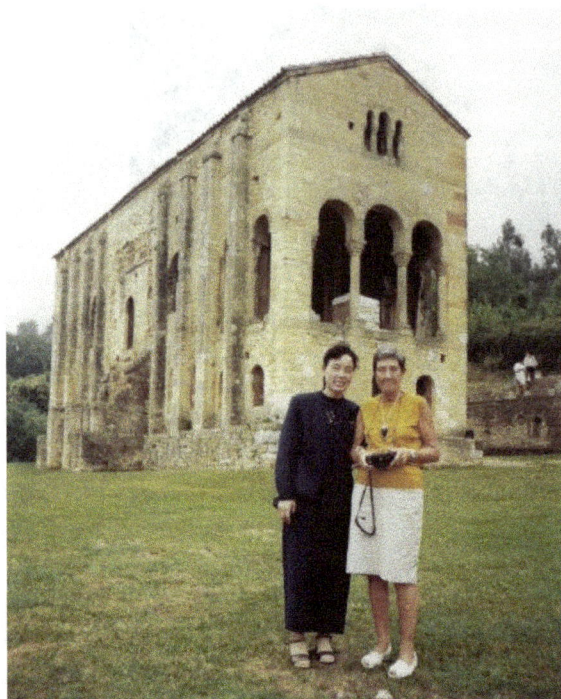

Fig. 1.34   With Madame Agnes Oberlin (right) in France in 2003.

month as a part of an international research project that involved myself (China) and Oberlin (France). Madame Oberlin is an expert in the use of transmission electron microscopy (abbreviated TEM) in the study of carbon materials such as carbon fibers. During this visit, she kindly taught me some TEM techniques for studying carbon materials. Thus, I was able to improve my research skills in this field. Moreover, understanding my research budget limitations, she suggested that I should use the fund meant for her accommodation and meals for the research instead. Professor Chengyang Wang 王成扬 of Tianjin University 天津大学, China, and Professor Cengmin Shen 沈曾民 of Beijing University of Chemical Engineering 北京化工大学, China (a leader in carbon fiber research), graciously allowed me to contribute to their funded research projects and use a part of their funding.

I particularly value the advice about the microscopy of carbon materials from Madame Agnes Oberlin, not only from her public presentation in my lab in Peking university, but also actual guidance in the TEM lab in both

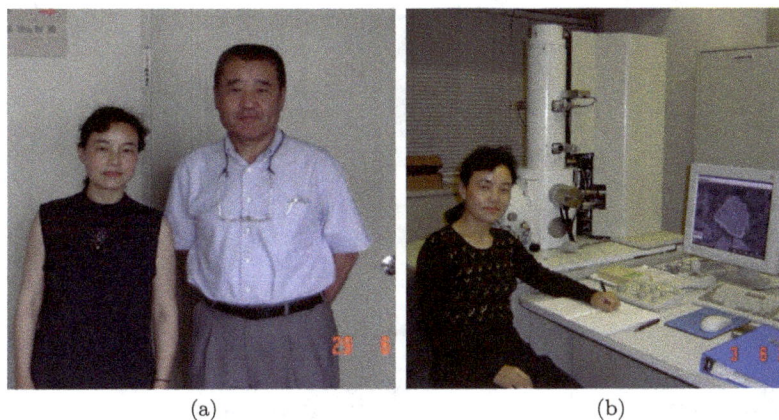

(a)   (b)

Fig. 1.35   (a) Receiving training in the laboratory of Professor Micheo Inagaki in Japan in 2003. (b) Working there on the porous materials by electron microscopy.

Beijing, China, and Mulhouse, Orleans, France. Professor Jean-Baptiste Donnet (Laboratoire de Chimie Physique, Ecole National Superieure de Chimie) in France gave me valuable knowledge about the resource of active carbon materials. I also received useful training about porous materials and advice on the giving of advice from Professor Micheo Inagaki of Japan (Fig. 1.35(a)) and also the heartwarming entertainment from Mrs. Inagaki (Itsuko). When a glass tube was broken during the transfer from another student (Fig. 1.35(b)), I was quite upset. In response, Professor Inagaki said, *"It is the cost of maturity and growth"*, and asked me not to worry about it. This experience helped me be able and willing to tolerate and afford the mistakes of students later. After his presentation in Peking university, he visited my lab, he gave me a nickname as the pioneer. It was revised as the pioneer of mineral materials, such as graphite mineral materials later by many friends later.

The U.S. scientists were typically attracted by scientists all over the world in international conferences. I'm a little timid and shy to join with them in the crowd. After some free face-to-face scientific discussions with some U.S. scientists, I did not feel a large communication gap from them. When I asked them scientific questions, I received not only the scientific answer, but sometimes also kind correction of my English language. With more communication and brave practice, my oral English was improved

Fig. 1.36   Visiting Professor D.D.L. Chung (right) of University at Buffalo, The State University of New York, in Buffalo, New York, U.S.A., and enjoying the "Endless soup & salad" in 2016.

and my knowledge was enlarged. Due to the fact that my second language is English, it seems not too difficult to work with the U.S. scientists.

As a visiting professor, I visited some universities and gave presentations in English in U.S. universities, such as University of Washington (Seattle), Michigan Technological University, University of Wisconsin (Madison), Columbia University, and the City College of New York. The research efficiency of the U.S. scientists was astonishing to me. Professor D.D.L. Chung often worked at the office even at midnight in The State University of New York at Buffalo. Due to the time shift between China and the U.S., it was understandable and enjoyable for me to be energetic for research discussion, but it was unbelievable to see such an energetic scientist till midnight. Her interests in science attracted me very much, like that amazing "Endless soup & salad" (Fig. 1.36). After half a month stay in her lab, I understood her fruitful works from her active, strict and efficient work. There was a kind of good frugal fashion there. An envelope was reused by many offices, sometime for 9 times. The scientists tried to spend every coin for the research work there and did very good efficient research.

I greatly value the training that I received in my university education and the special training that I received from the international experts. Continuous learning by reading and study by oneself is also necessary. Training and learning are indispensable for success in scientific research. Indeed, my father proudly said, *"My daughter got her permanent position in a top university after ten years of training beyond her bachelor's degree"*.

Fig. 1.37   With my Master thesis advisor, Professor Kelong Huang 黄克隆 (left) of Changchun College of Geology in 1988.

1.2.6.7 *The open topics of my Master and Ph.D. theses*

I completed my Master thesis in Changchun College of Geology (presently College of Earth Sciences, Jilin university) and Ph.D. thesis in China University of Geosciences (Beijing) with the help of professors who advised me well. The selection of a research topic was a critical first step for either thesis.

My Master thesis advisor, Professor Kelong Huang 黄克隆 (Fig. 1.37) of Changchun College of Geology was a specialist in the study of minerals by testing the effect of heating on the energy and mass of the minerals. He advised me to research on a particular type of porous mineral, but did not provide me with the scope, approach, information or specimens for conducting the study. I accepted the challenge, in spite of some anxiety about the feasibility of the topic. Then I proceeded to look for and obtain specimens of the mineral in the mountain. In addition, I read all the relevant literature that I could find in the library, including publications in Chinese, English, Russian and Japanese. Thus, I made a research plan, which also provided the thesis topics of two undergraduate students whom I supervised. The Master thesis resulted in my first two peer-reviewed professional journal papers, in addition to a review paper published in a national newspaper. The successful completion of the bachelor theses of the two undergraduate students contributed to my being able to obtain a Lecturer position in Xi'an University of Architecture and Technology.

Fig. 1.38   With my Ph.D. thesis advisor, Professor Xunruo Zhou 周珣若 (left) of China
University of Geosciences (Beijing) at my defense of my Ph.D. thesis in 1997.

I appreciate the experience that I gained from my Master thesis. The learning experience prepared me for my more fruitful research later. Since completion of the Master thesis, I became highly aware of the critical importance of selecting research topics. This thought has been occupying my mind so much that I often write unconsciously, "What is the topic?", when I have a pen in my hand; I do this even now. This continuous thought about a suitable topic helps me in finding topics that are scientifically and technologically meaningful and significant. This enabled me to compete successfully for research funding from the National Natural Science Foundation of China for numerous years.

I was fortunate to have met Professor Xunruo Zhou 周珣若 (Fig. 1.38), who received her Ph.D. degree from Moscow State University in the Soviet Union in the 1950s. She became my Ph.D. thesis advisor. With an open mind, she allowed me to select the research topic myself and gave me little limitation about the scope of the thesis. This freedom was wonderful, but it was also challenging. It gave me my first taste of what it takes to choose a topic for one's research.

I chose graphite for my Ph.D. thesis topic. Graphite (Fig. 1.39) is a kind of natural mineral. It is a crystalline form of the element carbon and is the most stable form of carbon. It is used for pencils and lubricants. Due to its electrical conductivity, graphite is also used for electrodes in batteries. One of my vice advisors had performed research on graphite, including

Fig. 1.39   Graphite, a mineral, which is used in pencils for writing.

the chemical modification of graphite with sulfuric acid (H_2SO_4). Graphite became my favorite mineral.

I made a research plan and successfully obtained a grant of about 10,000 Yuan (RMB) from a national laboratory to support the research. Thus, I started to perform experimental research on the chemical modification of graphite. The modification converts graphite to a compound known as a graphite intercalation compound (GIC). The compound is much more conductive than the unmodified graphite, so it is interesting from both scientific and technological viewpoints. I set up the experiment for performing the chemical modification and successfully prepared GIC from natural graphite. In addition, by rapid heating the GIC by using a welding torch, I prepared expanded graphite, which is a fluffy form of graphite obtained upon the extensive expansion of the graphite during heating. The expanded graphite is now a common starting point for the preparation of graphene. Graphene is a single layer of carbon (Fig. 1.40), in contrast of graphite, which has many layers stacked up. This research was fruitful and resulted in my first English paper, which was published in Carbon, the premier international journal in the field of carbon. With the publication of this paper and other papers in Chinese, I completed my Ph.D. thesis and received the award of the best Ph.D. thesis of China University of Geosciences, and I made good contributions to the field of graphite intercalation compound, particularly in relation to expanded graphite and the most concentrated form of GIC.

The success of my Ph.D. thesis reinforced my notion of the importance of letting the students have freedom in choosing their research topic.

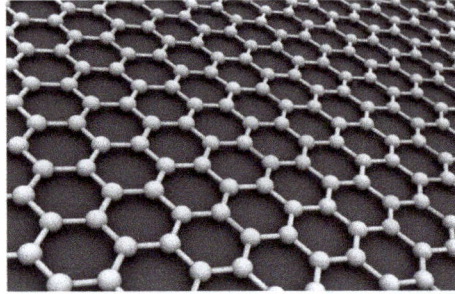

Fig. 1.40   Graphene (*https://news.flinders.edu.au/blog/2017/12/12/go-cleaner-efficient-batteries/graphene-lattice/*).

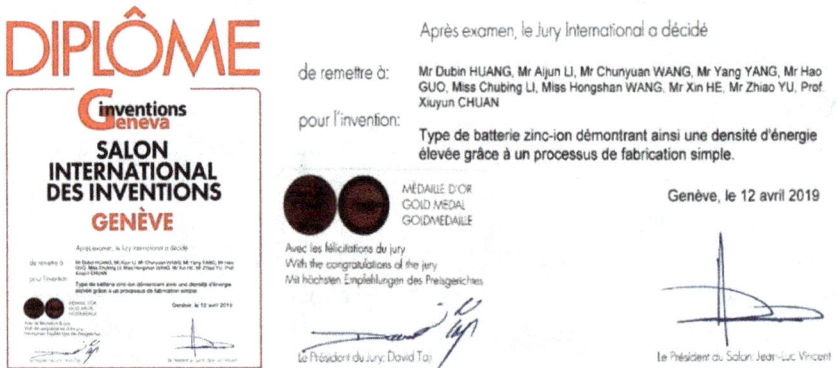

Fig. 1.41   The Inventions Geneva Prize awarded on April 12, 2019. The right part shows a magnified view of a part of the diploma.

Therefore, in my supervision of Ph.D. theses, I let the students pursue the research directions that interest them. Letting the students exercise their creativity is important for the development of the students into successful researchers. As a result, a special energy device (cell) was invented, thereby received the Inventions Geneva Prize (Fig. 1.41) on April 12, 2019, in addition to numerous other awards. Their theses were recognized as the best Ph.D. Theses of Peking University, and I was honored as the Best Advisor of Ph.D. Theses of Peking University.

Fig. 1.42   A youth acquiring the "scriptures" in the west.

1.2.7 *To acquire "scriptures" from the west*

Due to the severe difficulties that I faced in the early years of my research career, I longed for the good research conditions abroad, particularly in the western world. Therefore, with the hope of widening my horizon and enhancing my research, I took every opportunity to visit the laboratories of the active successful scientists in various countries, as well as the those of the great scientists of the past, such as Marie Curie (1867–1934) in Paris, France, Dmitri Ivanovich Mendelyeev (1834–1907) in St. Petersburg, Russia, German poet/dramatist Johann Wolfgang von Goethe (1744–1832) in Germany, etc., in order to enlarge my knowledge and open my eyes, as I tried to find the scientific way to become a great scientist (Fig. 1.42).

1.2.7.1 *Being inspired by Madame Curie*

When I was working in the research project that involved the collaboration of China and France in 2003, I visited Madame Marie Curie's laboratory located a humble courtyard at 33 Chateau de Loire in Paris, France (Fig. 1.43(a) and 1.43(b)). Professor Jean Yves Laval, the laboratory director, kindly guided me in the visit of Madame Curie's laboratory, and showed the equipment that Madame Curie used there (Fig. 1.43(c) and 1.43(d)). The laboratory was in a typical house with several typical rooms and was still used at the time of my visit. Madame Curie was the first woman to win a Nobel Prize, the first person and the only woman to win the Nobel Prize twice, and the only person to win the Nobel Prize in two scientific fields (Physics and Chemistry). Professor Laval also told me that three other scientists from her family also received the Nobel prize. These family members were her husband, her daughter, and her son-in-law. I was truly inspired.

No special public Memorial Hall for Marie Curie existed at that time, Professor Jean Yves Laval was proud of the great work of Madame Curie's family and was trying to establish the Curie Memorial Hall in another building nearby. I hope that the Memorial Hall will be realized in honor of Marie Curie.

Without enough funding or laboratory space, Madame Curie and her husband Pierre Curie persevered, and finally succeeded in extracting Radium after about 458 experiments over a period of four years. Due to her great concern for the welfare of mankind, she deliberately refrained from patenting the radium-isolation process, so that the scientific community could do research on the subject without hindrance.

Although her laboratory was very simple, with facilities and equipment that were not advanced, she worked diligently and made remarkable achievements. This points to the utmost importance of the mind (rather than the equipment or funding) in making research breakthroughs. Creativity and perseverance stem from the mind.

1.2.7.2 *Being inspired by Goethe*

There are numerous great German scientists who made significant scientific contributions. I was fortunate to receive financial support from a joint

Fig. 1.43   (a,b) Visiting the laboratory of Madame Curie in Paris, France. (c,d) Professor Jean Yves Laval, the laboratory director, showed me the equipment that Madame Curie used there.

research program between Germany and China, so I spent three months in the laboratory of German Chemist, Professor Peter Scharff, in Ilmenau Polytechnic University in Germany in 2009 (Fig. 1.44(a)). There I performed research (Fig. 1.44(b)), in addition to enjoying the discussion in the coffee breaks (Fig. 1.44(c)), and sharing a piece of calligraphy with Professor Peter Scharff (Fig. 1.44(d)). The stay enabled fruitful collaborative research, and helped me appreciate the strict scientific work habit of the German scientists. This habit was attractive to me and helped my subsequent research.

I was also inspired by German poet/dramatist Johann Wolfgang von Goethe (1744–1832). I visited his well-furnished apartment in Frankfurt, his university in Leipzig in central Germany, his office in the cute Weimar, where he was promoted to be a high official of the Weimar Republic and the mining engineer of an ore mine in Ilmenau, 100 kilometers from the Weimar capital.

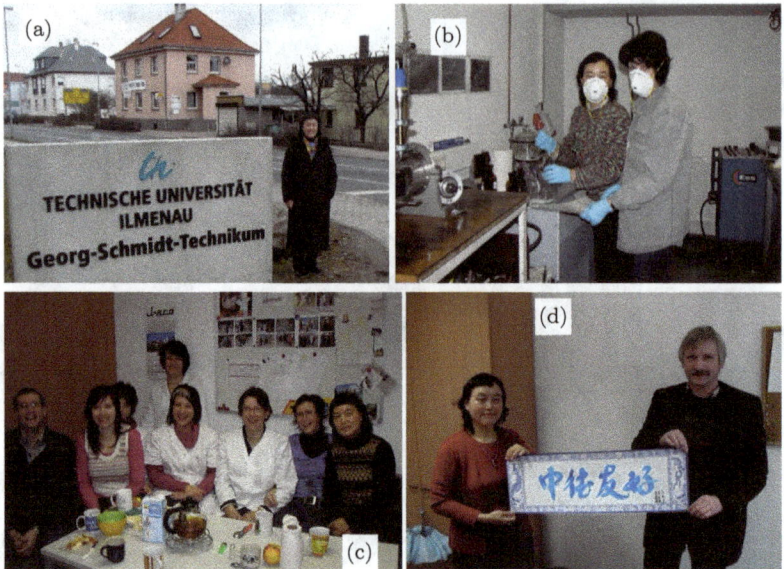

Fig. 1.44 Visiting Ilmenau Polytechnic University in Germany (a), working with Ms. Carmen Siegmund (right) at the laboratory (b), enjoying the coffee break with Prof. Uwe Ritter (left 1), and his group (c), and the calligraphy presentation to Professor Peter Scharff (right) (d).

In my spare time, I spent half a day to visit Goethe's cabin in the Thuringia forest (Fig. 1.45(a)). It is at the top of a mountain in a dense forest located in the south-eastern German state of Thuringia, where he wrote his famous poem in September 1780, when he was 31 years old in German language.

"WANDRERS NACHTLIED
Uber allen Gipfeln
Ist Ruh
In allen Wipfeln
Spurest du
Kaum einen Hauch;
Die vogelein Schweigen im Walde
Warte nur, balde
Ruhest du auch."

Fig. 1.45   (a) Goethe's cabin in Thuringia forest. (b) Goethe statue on a street in Ilmenau, Germany.

It was translated into many language, one translated version in English is popular there as bellow.

> *"**Wanderer's Night Song***
> *Hushed on the hills*
> *Is the breeze;*
> *Scarce by the zephyr,*
> *The trees*
> *Softly are pressed;*
> *The wood bird's asleep on the bough.*
> *Wait, then, and you*
> *Soon will find rest."*

I also enjoyed his lovely carriage, in which the coachman was dressed in his style in Weimar, in addition to learning about his lifestyle. I had a photo taken with the Goethe statue on a street in Ilmenau, Germany (Fig. 1.45(b)). As I was strolling on a narrow path in a park in Weimar, I even wrote a poem in his style.

Goethe was born in a rich family in Frankfurt. He graduated from the famous Leipzig University, and worked in a typical small town. He even lost his work some time, but he never gave up, and finally achieved his glory with his poems written in the vast Thuringian forest about 200 years ago. Thus, his humble cabin became the totem of many young students in

Fig. 1.46   (a) Dmitri Ivanovich Mendeleev. (b) His apartment and laboratory
(https://en.wikipedia.org/wiki/Dmitri_Mendeleev, public domain)

Germany. Due to his indomitable spirit and poetic life, whether he was rich
or poor, I also consider Goethe my totem. I also stepped on the snow that
was 30 cm thick to pray for the protection of the sages in the winter weather
of minus 20 °C.

1.2.7.3 *Being inspired by Mendelyeev*

Dmitry Ivanovich Mendelyeev (1834–1907) was a Russian chemist who
formulated the Periodic Table of the Elements, which is a tabular display
of the chemical elements (Fig. 1.46). The table shows periodic trends and
can be used to derive relationships between the various element properties.
The Periodic Table is the backbone of all of chemistry.

I have long known Mendelyeev from textbooks, but he seemed very
far from my reality. I had the opportunity to work in his university (Saint
Petersburg University, i.e., Saint Petersburg State University, Fig. 1.47(a))
in Saint Petersburg, Russia, for 3 months in 2016. Thus, I was able to visit
Mendeleev's apartment and laboratory there (Fig. 1.46(b), Fig. 1.47(a)).

Only some of the rooms in his apartment were used as his office and
lab. He shared his apartment with his family. Therefore, to me, the space
did not look like a real laboratory at all (Fig. 1.48(a) and Fig. 1.48(b)).
He kept in a room many samples of different types of rock and mineral.
Another room with many books, a table and a desk, was used as his
office. This means that he used some of the rooms of his home as his

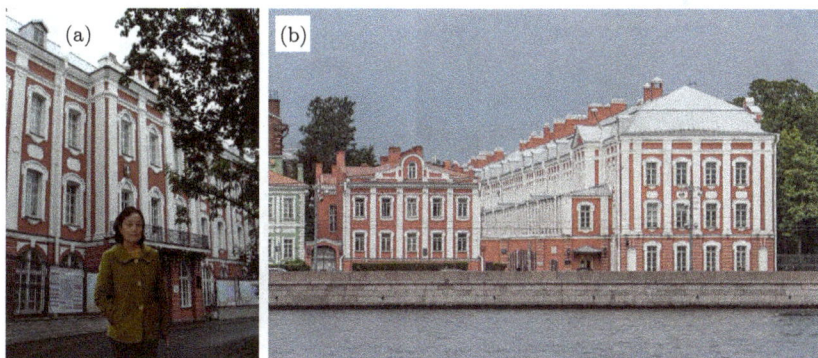

Fig. 1.47 Main administration building (a) of Saint Petersburg State University (b)
©A.Savin, WikiCommons
(*https://en.wikipedia.org/wiki/Saint_Petersburg_State_University*).

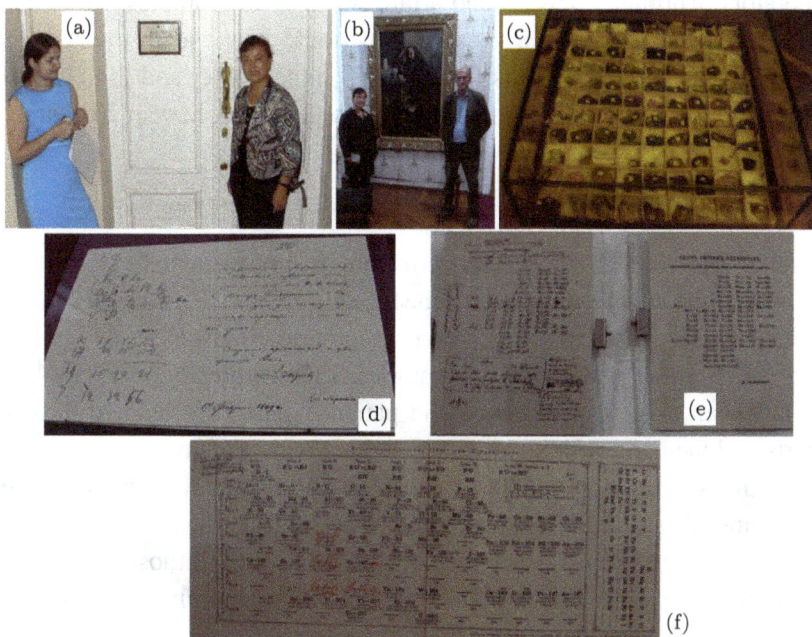

Fig. 1.48 Mendelyeev's laboratory in Saint-Petersburg State University on July 20, 2016,
the date of my visit with the guidance of (a) Ms. Yuliya Medvedeva and (b) Prof Valerii P
Tolstoy. (c) Mendelyeev's handwriting concerning numerous pieces of different types of
rock (d, e, f) Mendelyeev's sketch of the Periodic Table of the elements.

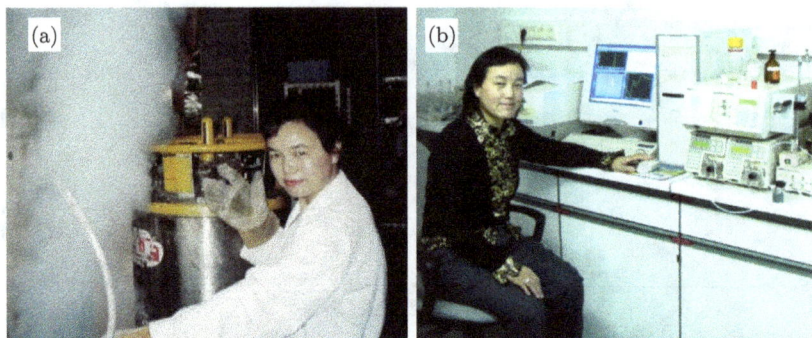

Fig. 1.49   Working in my laboratory. (a) Using liquid nitrogen to cool my samples. (b) Performing chemical analysis using various instruments.

laboratory. Through analyses and arrangement of the samples that he had collected (Fig. 1.48(c)), he set up the periodic law, and formulated the periodic table of the elements (Fig. 1.48(d), Fig. 1.48(e) and Fig. 1.48(f)) after numerous times of refinement over a period of more than 20 years.

1.2.7.4 *What I learned from the great scientists of the past*

In spite of my numerous trips to learn from the great scientists of the past, I failed to find any shortcut to become a great scientist. However, I came to understand the path to success. The path requires persistence and patience in research, rather than fancy laboratories. I felt fortunate, because I have always been used to persistence and patience, as I have learned through sports and the handwork at which my mother was very good.

Although the laboratories of Curie and Mendeleyev were not better than mine, these scientists made enormous contributions to science. My laboratory (Fig. 1.49) was actually not bad compared to those of Curie or Mendeleyev, so I became more confident of the possibility of becoming a good scientist, albeit not a great one.

Mendeleev had his own studio at home, where he used a collection of ore minerals for research. With the understanding of my husband and my occasional slight anxiety, I also did some experiments using the limited space at home.

The performance of scientific research does not require making a noise, but it requires the willingness to endure loneliness. Such endurance is particularly needed if one's research is way ahead of others. "I would rather gather mushrooms in the forest than live in a flashy society", as Mendeleev said. He did not care for honors or awards, but persistently pursues his research for more than 20 years. He did not receive the Nobel Prize, but his contribution to science is outstanding in the history of science and is beyond that of many winners of the Nobel Prize.

I also have been holding on to the attitude of persistency and patience in my decades of research. Madame Curie and Mendeleev continue to be powerful examples for me. It is important to persevere with one's goal in scientific research and continuously innovate ways to advance science for the benefit of society, whether the other scientists accept the work quickly or not, and regardless of one's stature or nationality.

1.2.7.5 *Financial management*

The amount of money a scientist earns is typically less than that earned by a successful business professional. Most scientists pursue science not because of money, but because of their lifelong interest in science. On the other hand, funding is necessary to support research. Both equipment and supplies need to be purchased. Particularly near the start of my scientific career, with the concurrence of my husband, I used a part of my personal saving to support my research. This investment turned out to be fruitful, as it enabled my research to progress toward my promotion to the rank of Associate Professor and later Professor.

Wealth is often viewed in today's society as an indication of success. However, for a scientist, success is indicated by one's impact on science and satisfaction is derived from the pursuit of one's interest and the consequent breakthrough in research. I am fortunate to have been able to pursue research in the topics that interest me greatly, particularly research concerning carbon materials. In spite of my countless failures in research and the constant challenge of searching for suitable research topics, I never gave up. Being able to sustain the research for over 20 years also played an important role in the advancement of my career.

Due to the quality of my research and my promotion to the rank of Professor, I was honored by being described in terms of the Chinese proverb, *"Real gold does not need plating, clear water does not need perfume,* 真金不镀，清水无香" by Professor Jing-Yin Li 李景荫, a famous old scientist, who wrote the proverb using his own calligraphy. This proverb means that gold is shiny enough even without plating, and clear water is delicious without perfume.

1.2.8 *Continuous learning*

Continuous learning is critically important, whether one is a student or a professional. This is because knowledge is limitless, even within one's area of expertise. Besides, one's interest should not be limited to one area, as having knowledge in multiple areas enhances one's chance of making breakthroughs. For example, if one's major is inorganic chemistry, one should study organic chemistry as well. By restricting one's interest to a single narrow area, one is missing out on the broader excitement of science, including aspects of the beauty and applications of science that may be valuable to one's research. Continuous learning is needed not only for research, but is needed in any profession, even farming. However, continuous learning requires humility and an appetite for learning. In addition, learning outside one's area of expertise requires courage and determination, as it is like leaving one's comfort zone. Broad-based learning in the university helps the learning outside one's comfort zone during one's subsequent career. It takes an effort to learn, but the reward can be great. By learning, one's horizon is widened, thus increasing the chance of making breakthroughs. In the long run over one's career, learning typically involves self-study, i.e., reading books and literature in the absence of a teacher. Self-study is easier if one is broad in the fundamental education, such as the basic education in physics and chemistry. Self-study may be supplemented by attending conferences, lectures and courses. For example, if one's research is on a particular material, one should attend seminars not only on this particular material, but also those on other materials. Even today, I am still learning. The more I learn, the wider is my horizon and the more exciting my research and teaching become.

Fig. 1.50   The logo of Peking University and its meaning. (a) The university logo. (b) Meaning 1, one (at the bottom) supporting the two (at the top). (c) Meaning 2, one (at the bottom) growing up through hard work to pass through the gap between the two.

I also learned a number of languages, including English, Japanese, French, German, Korean, Russian and Portuguese. Only two languages (Russian and Japanese) were learned when I was a graduate student. The rest was learned after I had worked, even after I became a professor. With more foreign languages studied, the world became wider and the research work became more interesting and enjoyable. Knowing the languages also helped my greatly when I visited various countries, including Japan, France, Germany, Brazil, Spain, Korea, Russian, Italian, Austria, Czech, the United States, etc.

In relation to raising a family, I learned how to give birth to, feed, take care of and educate young children. In relation to my teaching work, I learned how to teach, how to organize classes, how to communicate with others, and how to teach the youth about science. Studying is needed for all aspects of life and helps one know more about oneself as well as others. The knowledge gained enables one to go through the hurdles of life more smoothly. It also enhances one's appreciation of the world and its cultures and events. It is never too late to learn.

Life can be difficult, but it is beautiful, if it is well lived. Science is wonderful, but the road to scientific success can be challenging. The logo of Peking University consists of two Chinese characters, namely north (meaning Peking) at the top and large (meaning university) at the bottom (Fig. 1.50(a)) in Chinese. The combination of the two characters depicts three persons separated by narrow gaps, with the upper two persons being side by side but near each other, and the lower person looking like a hercules. In my opinion, this logo resembles a professor (at the bottom) supporting two young students (at the top) (Fig. 1.50(b)) through hard work in teaching and self-improvement. Alternatively, the logo can be considered to depict a youth (at the bottom) growing up in stature through hard work to pass through the gap between the two professors (at the top) (Fig. 1.50(c)). The gap between the two professors (at the top) may be considered a route to success and also a ship heading to the sea. Through my work experience of more than 30 years, I am convinced that hard work is essential for success, regardless of the age or career level of the person.

1.2.9 *Concluding remarks*

The road to scientific success is not easy, as it takes diligence, creativity, sustained work and continuous broad-based learning. The quest for knowledge is a lifelong endeavor that brings joy and satisfaction beyond what money can provide. Scientific success is not measured by awards or job titles, but it is measured by the impact that one gives to science, technology and society. Science is the foundation of any technology, whether electronic, computer, automobile, aerospace, energy, environmental or medical technologies, but its advancement requires dedicated work that may take decades.

It was a miracle for an underprivileged girl from a primitive cave dwelling in the poorest part of rural China to become a scientist and a professor in a top university. For a girl with inadequate food, clothing, lighting or learning opportunity in a male-dominated village, it was a dream beyond imagination.

I am grateful to my parents for their sacrifice and encouragement, which were crucial for making my quest for science possible. I am thankful to my elder sister, who gave me the shirt used in her wedding when I left the village for the university. I am also thankful to my young brother and sister, who saved the family money for my lodging and living expenses in the university, in addition to preparing the deep-fried dough sticks to be sold. The family is very important. One must cherish every member of one's family. I also appreciate my husband for his constant support.

I am grateful to my mentors in China and other countries, particularly my Ph.D. dissertation advisor and the international carbon scientists, who helped me obtain research funding and taught me valuable research techniques. I am also grateful to Changchun College of Geology for providing me scholarships and sports-related food assistance when I was a student. Moreover, I am grateful to Peking University for giving me a free rein in performing research as a professor. Health is precious. Always take care of your health.

Whoever you are, wherever you came from, it is possible for your dream to come true. Keep working toward your goal and never give up. Be thorough in your major field of study. Be brave to venture into fields outside your comfort zone. Enjoy the gaining of knowledge and the widening of your horizon. Learn various cultures and interact with people in various parts of the world. At all times, take care of your family, particularly your parents, in addition to making and maintaining your friends. Whatever is your professional goal, be sure that you do everything in an ethical fashion, regardless of the difficulties that you may face. Life is short. Enjoy it and make the best use of it.

Chapter 2

Zulkhair Mansurov – from Kazakhstan to the world through combustion science

2.1 Introduction by the Editor

The Soviet Union (the Union of Soviet Socialist Republics or USSR) was a federal socialist state in Northern Eurasia). The USSR started in 1922 following the Russian Revolution that ended the monarchy of Tsar Nicholas II. It existed till 1991 as a one-party state governed by the Communist Party. Moscow was its capital. However, in 1991, the USSR was dissolved.

Kazakhstan (Republic of Kazakhstan) was a constituent (union) republic of the USSR in 1920–1991. However, along with the dissolution of the USSR, it became independent on December 16, 1991.

Kazakhstan is the largest country in Central Asia (Fig. 2.1). It is an essentially landlocked country. Its neighboring countries are Russia, China, and the Central Asian countries of Kyrgyzstan, Uzbekistan, and Turkmenistan (Fig. 2.2). The culture of Kazakhstan was influenced by the ancient Silk Road (Fig. 2.3), Nomadic lifestyle, and Russia. In particular, the Silk Road was a network of trade routes connecting the East and West. It greatly facilitated the economic, cultural, political, and religious interactions between the East and West from the 2nd century BCE to the 18th century.

Kazakhstan has large reserves of oil, natural gas and coal, and is a net energy exporter. A hydrocarbon refers to a compound of hydrogen and carbon. The main components of oil and natural gas are hydrocarbons. Oil (petroleum) is a liquid mixture of hydrocarbons present in certain rock strata. The extraction and refining of oil results in gasoline, kerosene, diesel

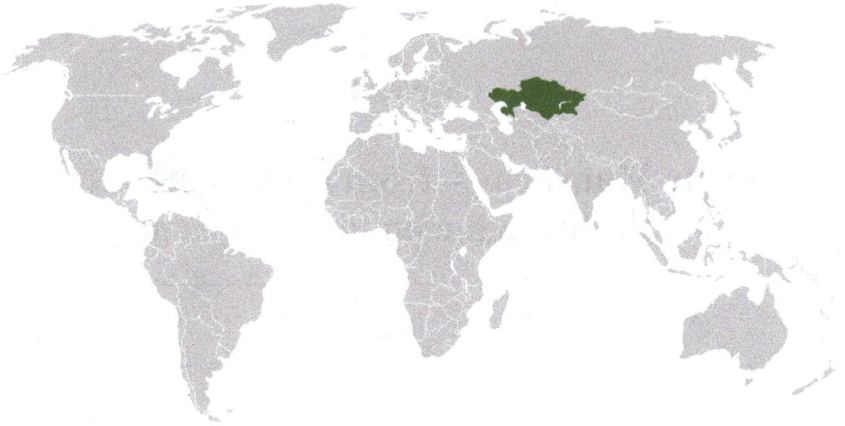

Fig. 2.1 Kazakhstan is the largest country in Central Asia. (https://en.wikipedia.org/wiki
/History_of_Kazakhstan, public domain)

Fig. 2.2 Kazakhstan and its neighboring countries. (https://hif.wikipedia.org/wiki/file:K
azakhstan_political_map_2000.jpg, public domain)

Fig. 2.3   An ancient silk road in the Valley of the Castles (a part of Charyn Canyon) in Kazakhstan. (https://en.wikipedia.org/wiki/Charyn_Canyon, attributed to Jonas Satkauskas)

oil, etc., which are widely used as fuels. Natural gas (also simply called gas) is a naturally occurring hydrocarbon gas mixture consisting mainly of methane (CH_4 molecules). Coal is a combustible black or brownish-black sedimentary rock that mainly comprises solid carbon. Energy is obtained from oil, natural gas or coal by combustion, which means burning. Combustion is a very important process in our daily lives. For example, the combustion of gasoline is used to run our cars, and the combustion of natural gas is used to heat our houses.

Combustion involves a chemical reaction that releases energy. An example of a combustion reaction is shown below for the combustion of methane (CH_4). In the combustion reaction illustrated in Fig. 2.4, methane reacts with oxygen to form carbon dioxide and water. In general, the combustion reaction is fed by a fuel, oxygen and heat (Fig. 2.5). The fuel is a substance that can burn. Oxygen acts as an oxidizing agent that facilitates the combustion. The heat is needed to start the burning. The fuel is in the form of molecules (such as methane) that contain carbon. The reaction

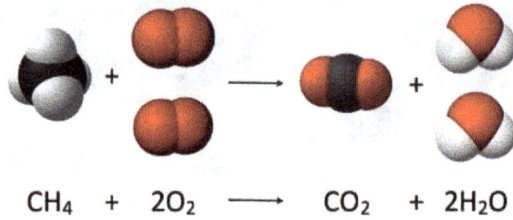

$$CH_4 + 2O_2 \longrightarrow CO_2 + 2H_2O$$

Fig. 2.4 Combustion of methane. (https://en.wikipedia.org/wiki/Combustion, public domain)

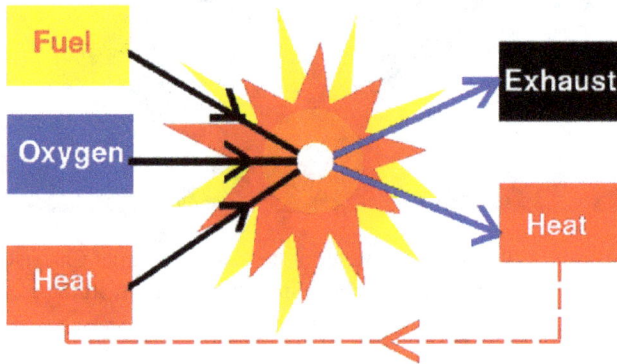

Fig. 2.5 The combustion reaction in general. (https://wright.nasa.gov/airplane/combst1. html, public domain)

generates heat and exhaust (such as carbon dioxide and water in case of the combustion of methane, Fig. 2.4). The reaction is exothermic, i.e., it produces heat, which can be partly used to feed the reaction further, so that the reaction is sustained, as illustrated in Fig. 2.5. Combustion is typically not complete, so there are products that are unburned or partially oxidized. An example of such as product is soot (impure carbon particles) (Fig. 2.6). The formation of soot occurs in all hydrocarbon flames. The soot makes the flame luminous and nontransparent.

Pyrolysis differs from combustion in that it refers to the thermal decomposition of materials at elevated temperatures in an inert atmosphere. Pyrolysis of organic substances produces volatile products, in addition to leaving a solid residue that is relatively rich in carbon. This residue is known as char. In case that the residue is essentially carbon, the pyrolysis is known

Fig. 2.6   Smoke containing soot, as emitted by a truck that burns diesel fuel without a filter for the exhaust. (https://en.wikipedia.org/wiki/Soot, public domain)

as carbonization. Pyrolysis is considered as the first step in the process of combustion.

Professor Zulkhair Mansurov (June 26, 1946-) of Kazakhstan is an international expert in combustion science. He has overcome severe difficulties related to the conditions in his country. Through his perseverance and dedicated research, he has contributed extensively to the field of combustion science, in addition to elevating greatly the level of scientific research in Kazakhstan.

A nanomaterial refers to a material that has particles or constituents that have nanoscale dimensions (i.e., nanometer scale, with $1\,nm = 10^{-9}\,m$). An example of a nanomaterial is the carbon nanotube, which is in the form of a tube having a cylindrical wall and a hollow channel at the center of the tube along its axis (Fig. 2.7). In 1996, the Nobel Prize in Chemistry was awarded to Richard E. Smalley and Robert F. Curl Jr. of Rice University in Houston, Texas, U.S.A., and Harold W. Kroto of the University of

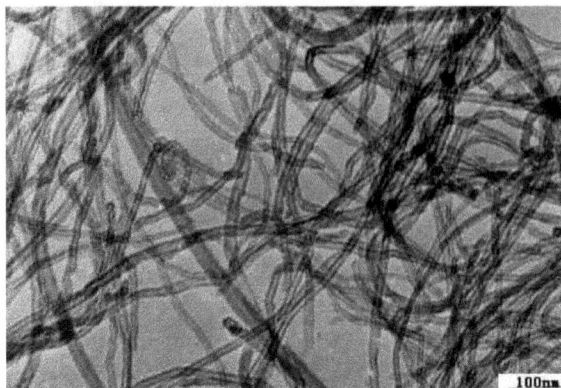

Fig. 2.7 Carbon nanotubes viewed by transmission electron microscopy. The scale bar indicates 100 nanometers (1 nanometer $= 10^{-10}$ meter). Each nanotube has outer diameter 10–20 nm, inner diameter 3–5 nm, and length 10–30 μm. (https://www.cheaptub es.com/product/oh-functionalized-graphitized-multi-walled-carbon-nanotubes-10-20n m/?gclid=COWXn-X_v88CFYlbhgodOTcMAQ, public domain).

Sussex in Brighton, United Kingdom, for their discovery of the fullerene, which is the C_{60} molecule (a molecule with 60 carbon atoms and no other atom).

The science of nanomaterials has advanced greatly over the last 20 years. Combustion is now considered as a method for the synthesis of nanomaterials, such as fullerene, due to the research of Professor Mansurov (Fig. 2.8). Nanomaterials in the form of carbons are particularly important for the purification of water and blood. In 2018, Professor Mansurov received a UNESCO Medal for the development of nanoscience and nanotechnology. UNESCO stands for the United Nations Educational, Scientific and Cultural Organization. It is an agency of the United Nations and is aimed at contributing to the building of peace, the eradication of poverty, sustainable development and intercultural dialogue through education, the sciences, culture, communication and information. In addition, Professor Mansurov received the State Prize of Kazakhstan in 1992. In 2002, Professor Mansurov and his research associates received a Diploma for the discovery of "Phenomenon of low-temperature cool-flame soot formation", as issued by the Russian Academy of Natural Sciences.

Fig. 2.8   Professor Zulkhair Mansurov. (https://www.researchgate.net/profile/Zulkhair_
Mansurov)

2.2 Life experience as described by Professor Zulkhair Mansurov

2.2.1 *My parents and World War II*

My father, Aimukhamet Kopezhanovich (1917–2000), was born during the period of grandiose tragic and vast changes in Russia. The turmoil in the heart of Russia between the two revolutions echoed on the outskirts of the empire with a vengeance and effects that are most often negatively (Fig. 2.9). Such grandiose changes broke the character and lifestyle of not one person, but the society as a whole, changing the value and cultural orientations of a person.

The 100th anniversary (birthday) of my father in 2017 was a significant event for my large family. Father lived a very complex, difficult, but interesting life. He was a man, a hero and a tireless worker of his time. His life was a reflection of the history of Kazakhstan.

My father was always grateful to his parents. He respected them for the fact that, under the difficult conditions, they were able to educate him, raise

Fig. 2.9 The Russian Revolution that occurred after World War I and resulted in the end of the Russian monarchy in 1917. The photo shows soldiers demonstrating in Russia in Feb. 1917. In 1920, Kazakhstan became a part of U.S.S.R. (https://en.wikipedia.org/wiki/Russian_Revolution, public domain)

him as a patriot that preserves the traditions of his people and becomes a true internationalist. My father loved and was proud of his parents. In his memoirs, he wrote that his father (Kopezhan) took him to school when he was seven years old. And my father ordered us not to change our last name (Mansur).

History teaches that wars are inevitable, as they occur to resolve territorial and political problems. In response to the call of Joseph Stalin (1878–1953, Soviet politician who led the Soviet Union from the mid-1920s until 1953 as the general secretary of the Communist Party of the Soviet Union and premier of the Soviet Union), Kazakhstan participated in the European resistance movement during World War II. Thus, war befell my father (Fig. 2.10). It was so cruel that Father did not like to talk about the war. However, he remembered Victory Day. It was interesting for us to listen to him. He was a paratrooper. Even at the present time, landing troops are

Fig. 2.10   My father, Mansurov Aimukhamet Kopezhanovich (in the center), with his fellow soldiers, in 1944.

considered to be elite, as they are quick reaction troops. He remembered that when they were prepared in the first months of hand-to-hand combat, the instructors said that the enemy needed to be shot down with one blow, as there might not be a second.

After suffering from a double wound, he miraculously survived. The Army services sent to his mother news about his son's death, but his mother hid it, without telling anyone. My grandmother believed that her son was alive and waited for him. Later, he was taken to the hospital in Yaroslavl (a Russian city located northeast of Moscow). After a few months, he was discharged from the hospital and returned home.

2.2.2 *My upbringing*

I was born on June 26, 1946 in the small town of Sarkand in the Taldy-Kurgan region of Kazakhstan. But now it is Almaty region (Kazakhstan's largest metropolis). My father Aimukhamet Kopezhanovich (Fig. 2.10) fought at the front and was repeatedly awarded for his military work.

Fig. 2.11 Father Aimukhamet Kopezhanovich (sitting first on the right), myself in the arms of my uncle Orazbay (an elder brother of my father), my grandmother Kasipa. Close to her from left to right standing: Musa, Talgat, Maria, Token (they are children of my father's elder brother Satibay); the woman with a scarf on her head – my mother Maken Tursynbekovna (standing second from the right), 1946.

My mother Maken Tursunbekovna (1924–2008) (Fig. 2.11) gave birth to seven children. I have 3 brothers and 3 sisters. I am the oldest. Our parents raised us in respect and love to one another and to our family, relative, friends and colleagues. My parents emphasized the preservation of family traditions. I grew up with love to our native land and respect to our ancestors.

Father always chose names for his children and grandchildren. It was a special ritual. When I was born, he decided to call me the name of his fellow soldier Zulkhair Baiseitov, who worked then in the Karaganda region (a region of Kazakhstan), but could not immediately contact him in order to get his consent. And only 40 days after my birth, I got my name.

Father paid much attention to physical education. He took cold baths. In the winter, he liked to walk barefoot on the snow. I remember how my father awakened us at 7 a.m. once a week, in order for us to swim in the cold river om the countryside. My father wanted to prepare my brother and me for independent living, so he regularly sent me to pioneer camps.

Fig. 2.12 My high school graduation class in Talgar. I am the fourth from left to right on the second line from the top, 1963.

After completing the 9th grade in Talgar (Fig. 2.12), I hiked by foot with the school group to Issyk-Kul (one of the regions of Kyrgyzstan, a country bordered by Kazakhstan to the north, Fig. 2.2) and back. During the winter holidays, I made a six-day 106-km ski trip from Talgar to Chilik. Both town are in Almaty Region, southeastern Kazakhstan. Father and mother tried to give us comprehensive education, with the purpose of developing us into harmonious people with a broad outlook. Therefore, my younger brother Tair (1948-) and I were sent to a music school in Panfilov (a district of Almaty Region in Kazakhstan). There we learned to play the button accordion.

Father often quoted methods of education - Socrates, Aristotle and others. But in practice, he chose the simplest and most effective method. He used his personal example both in work and in everyday life. As long as I can remember, my father got up early, his working day was filled until late in the evening. Through his life and work, he showed us an example of responsibility, an active life position in front of a collective society, and the need to be well-versed in the chronicle of life events in the country and the world. He always said that one must do good and not evil. He also said, "If you can't help with something, don't promise, but if you promised, do it."

My mother died in 2008. In the 2009 memorial book about her, I wrote the following.

Find the exact words to express all your son's love. Therefore, I recall the lines of one of the famous songs of Bulat Okudzhava (1924–1997, a Soviet and Russian poet, writer, musician, novelist, and singer-songwriter of Georgian-Armenian ancestry):

> *There are so few real people!*
> *All of you are lying that their age has come.*
> *Count both honestly and strictly,*
> *How much will be for each quarter.*
> *There are very few real people.*
> *On the planet - absolutely nonsense.*
> *To Russia - one of my mothers*

So my mother was a real, reliable, kind and caring person, not only for her children, her grandchildren and our entire family, but also for many who knew and loved her. I am immensely grateful to my parents (Fig. 2.13) for the real upbringing we received, for the life school, and for the kindness and cordiality that support us.

2.2.3 *My dear wife and her untimely death*

In 1963, Raushan Magzumovna (1946–2006) and I entered the Chemistry Department of Kazakh State University (KazSU, currently known as Al-Farabi Kazakh National University) in Almaty in 1965 (Fig. 2.14). We studied in the same group, and got married before graduation from the university in 1968 (Fig. 2.15).

Parents, children, grandchildren and other relatives constitute the most important refuge in our difficult lives. The difficulty is especially acute when, by fate or natural causes, relatives die. The tragedy of an inexorable coincidence turned out for our family in relation to the untimely death of

Fig. 2.13 My parents: Aimukhamet Kopezhanovich and Maken Tursynbekovna Mansurov, 1987.

my beloved wife, Raushan after a serious illness. She was wife, mother, grandmother, kind sister, and guardian of our family hearth.

Raushan was our support in everything (Fig. 2.16). She graduated from the university in 1968, pursued an excellent scientific career, became a Doctor of Chemical Sciences and a Professor at the al-Farabi Kazakh National University University. She trained many students, and contributed to the development of domestic science and higher education.

I and Raushan raised two sons. They are Kadyr (1969–1995) and Batyr (1972–2018). The upbringing of children and their development make up a very important and interesting period in the life of parents. I remember with warmth how they grew up, went to kindergarten, and studied in school. And here, of course, the merit goes to Raushan, since most of the care about raising children fell on her shoulders. Raushan read many books to them. They together learned poems and songs.

My elder son Kadyr graduated from the Moscow Institute of Physics and Technology (a Russian university). Unfortunately, on the way back to Kazakhstan in 1995, he died in a traffic accident in Moscow. It was a huge

Fig. 2.14 The main building of Kazakh State University in Almaty, Kazakhstan, 1934. One year after Kazakhstan's 1990 Declaration of Independence, the name of the university was changed to Al-Farabi Kazakh State University. The university currently has 20,000 students (of whom 4,000 are graduate students and 16,000 are undergraduate students) and more than 2,500 faculty members. (https://en.wikipedia.org/wiki/Al-Farabi_Kazakh_Na tional_University, public domain)

Fig. 2.15 Raushan and I on July 6, 1968. Invitation card for our wedding in the Kazakh language.

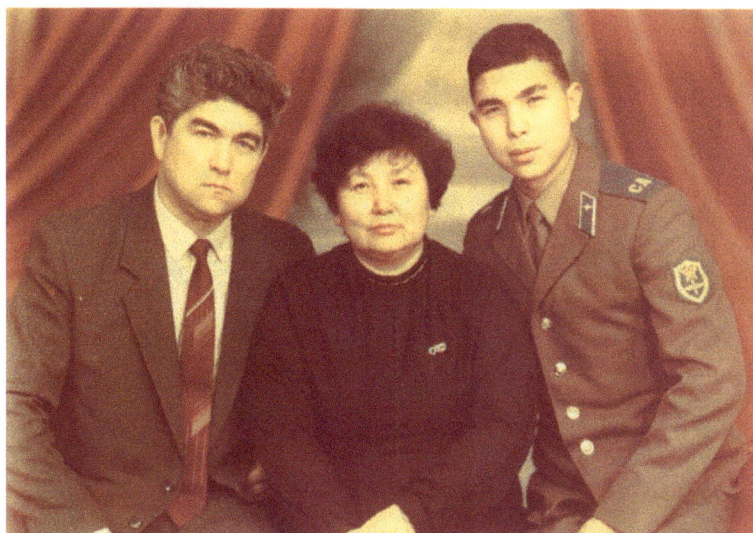

Fig. 2.16   From left to the right: I, my wife Raushan Magzumovna and our son Kadyr in Ivano-Frankivsk (a historic city located in Western Ukraine), 1989.

blow for my family. The impact on Raushan was so great that she developed heart problems. In 2006, she died.

My younger son Batyr graduated from the Physics Department of Al-Farabi Kazakh State University. He worked as a senior teacher at the Department of Solid State Physics. Unfortunately, he passed away in 2019 after a stroke. Of course, I have to say kind words about Madina, the wife of Batyr. She is the Head of the Department of Artificial Intelligence and Big Data in Al-Farabi Kazakh National University. Being calm and caring, Madina became the darling of the big family of the Mansurovs. She gave us great joy - granddaughter Benazir (1994-) and grandson Nur-Kadyr (1997-) (Fig. 2.17 and 2.18).

The grief resulting from the deaths of my wife and both sons was indescribably intense. However, my family, relatives and friends gave me the needed support, so that I was able to overcome the grief and continue with my work. Working hard also helped me survive in those difficult moments of my life.

Fig. 2.17 My family. From left to the right. (Standing): son Batyr, daughter-in-law Madina. (Sitting): wife Raushan, granddaughter Benazir, grandson Nur-Kadyr and I, 2005.

Fig. 2.18 With grandson Nur-Kadyr, 2016.

2.2.4 *My educational and scientific life*

After the collapse of the USSR, the economy was bad in Kazakhstan. However, science began to be funded, doctoral programs of study started, and international scientific cooperation became increasingly feasible. I am thankful for the opening of scientific opportunities in Kazakhstan.

I graduated from tenth grade standard school in Talgar in 1963. I still remember my teachers from secondary school No. 1 with great gratitude.

In subsequent years, my dad sympathized with my choice of the faculty of Chemistry of Kazakh State University (KazSU), because I was convinced that KazSU was the flagship of education in Kazakhstan, and Chemistry at that time was a popular and promising branch of science and technology. The main testament of my father, which he gave, sounded simple: "To be honest and a good specialist."

I read a lot about travelers and geologists, so I wanted to join the Geological Faculty of the Polytechnic Institute. However, my mother was against it. She said, "No need to wander far away." My father was not opposed to this. He said that this was a good male profession. At that time, Nikita Khrushchev (1894–1971), General Secretary of the Central Committee of the Communist Party of the Soviet Union, at one of the party forums, expanded the wording of Vladimir Lenin (1870–1924, Head of Soviet Union in 1922–1924), "*Communism is the Soviet government, plus the electrification of the whole country, plus farms*" by adding "*plus the chemicalization of the national economy.*" This amended slogan brought about a clothing revolution (1950s and 1960s in the Soviet Union) that resulted from the chemistry related to synthetic fabrics. The headline "*Plus chemicalization*" went roaming the newspapers. Thus, I did not become a geologist, although this was my youthful romantic dream, but joined the Chemical Department of Kazakh State University (also named after Soviet party and state leader S.M. Kirov, 1886–1934). This step turned out to determine my entire future fate as a person, scientist and teacher.

In the third year of my study in Kazakh State University, it was necessary to choose the department to enter. At that time, the Department of High Molecular Compounds (HMC) was under the supervision of S.R. Rafikov, and B.A. Zhubanov (a young Doctor of Chemical Sciences) started working there. Polymers were fashionable; they were widely introduced

Fig. 2.19   With Prof. G.I. Ksandopulo, 1983.

into products such as nylon shirts, bologna raincoats, etc. The competition at the Department of High Molecular Compounds was intense, but I thought of going there. However, I was invited by G.I. Ksandopulo (Fig. 2.19) of Department of Physical Chemistry to his laboratory and was offered to work on an EPR-spectrometer (EPR standing for electron paramagnetic resonance) for the purpose of studying atoms, free radicals, and reactive particles, which play an important role in combustion processes. After careful consideration, there was no doubt that I had to join Department of Physical Chemistry.

By this time (since 1966), with the support of Academician M.I. Usanovich, G.I. Ksandopulo (or, as we said, G.I.) had conducted research in the field of the chemical physics of combustion processes using ultramodern physical methods, namely mass spectrometry and EPR spectroscopy. As the Head of "career guidance work", Georgy Ivanovich outlined clearly the prospects of the scientific direction.

After graduating from the Chemical Faculty in 1968, I began my work as a trainee-researcher at the Department of Physical Chemistry of Kazakh State University. In 1969 I was called into the Soviet Army, served in its ranks for a year and returned for postgraduate study at Physical Chemistry Department in 1970.

In 1973, I defended my candidate dissertation. I published an abstract and presented it to my father and mother. Naturally, this is a dissertation abstract on the topic "Kinetics of the interaction of hydrogen atoms with inhibitors". Father looked through the abstract, and asked, "And why, on the first page, it is not indicated that the work is carried out in accordance with the decisions of the Congresses and Resolutions of the CPSU Central Committee? That is what humanities usually do." I replied that this was Chemical Physics, and this is in accordance with my supervisor, G.I. Ksandopulo, Institute of Combustion Problems, Kazakh National University. I graduated in 1973 with a Doctor of Philosophy degree from Kazakh State National University. My dissertation was devoted to the development of the "contact time" method for determining the reaction rate constants of hydrogen atoms with various inhibitors of combustion processes using Electron Spin Resonance (EPR) spectroscopy. This experimental technique is for studying the structure of a material through observing the effect of a magnetic field on the electrons in the material. My parents and relatives attended my dissertation defense, which was successful. To celebrate, my younger brother Tair organized a good banquet in a restaurant in Kok-Tobe (a mountain in Almaty, Kazakhstan).

Tair is a politician, being the General Secretary of the Eurasian Economic Community, Ambassador Extraordinary, Moscow (Fig. 2.20 and 2.21). He has received numerous awards, including the Russian Federation Presidential and Belorussiya Orders of Honour, Order of Friendship (Kazakhstan), and Order of Nazarbaev.

My first scientific trip was in 1969. I prepared a report on the kinetics of hydrogen atoms interaction with combustion inhibitors, as studied using EPR spectroscopy (Fig. 2.22). I presented it at the 25th anniversary conference at Kazan State University in Kazan, Russia. The EPR phenomenon was discovered by an outstanding scientist academician E.K. Zavoysky (1907–1976) in 1944. It was a celebration of science with participation of its luminaries - Nobel laureates, famous professors, and prominent scientists. Since that time, I have been keeping diaries of my scientific and business trips, the number of which has now exceeded 160.

From 1979 to 1987, while working as the Head of Physicochemical Research Methods Laboratory, G.I. Ksandopulo actively participated in work of the department and taught special courses on "Combustion Theory" and "Kinetics of Gas-Phase Reactions". I would like to note the great work of the department staff under his guidance.

Fig. 2.20　With my young brother Tair Mansurov (General Secretary of the Eurasian Economic Community, Ambassador Extraordinary, Moscow) in the Embassy of Kazakhstan in Russia, 1996.

Although Kazakh is my mother tongue, I started learning Russian when I was about 2-3 years old. My family spoke Kazakh and Russian at home. I started studying English on my own in 1966, when I was 20 years old. In 1976, the university organized English courses for teachers and researchers, as taught by Zavgorodnaya Tamara Yakovlevna. Classes were held twice a week from 8.00 to 9.45 am. The language study group included Doctors of Science and Heads of departments. At that time, I worked as a junior Research Fellow. I remember that, in the first lesson, everyone was supposed to introduce themselves to the class in English. Besides me, all scientists had already been abroad and could easily introduce themselves and explain themselves in English: "My name is ... (Sultangazin, Zhubanov, Peretyatkin, Ksandopulo, Zhaksykbaev, Akanaev, etc.). I am Professor of Chemical Sciences." I was the last to introduce myself. My pronunciation

Fig. 2.21   My young brother Tair Mansurov (right) meeting with Russian President Vladimir Putin (1952-, left), 2008. (http://archive.premier.gov.ru/eng/events/news/1458/)

was so bad that she said, "Young man, in my opinion, you have come to the wrong place." I asked if I could stay.

My senior colleagues held high positions and often missed the English classes. In contrast, I attended the classes regularly. I visited the library often. I had the habit of writing words on library cards, with a word written on one side and tha translatino written on the reverse side. I had many such cards. In the morning, when I was on the No. 22 bus on my way to work, I went to the end of the bus, where I learned words and phrases. In the English lessons, we worked out three books by English authors, namely "Robinson Crusoe" by Daniel Defoe, "Moonstone" by Wilkie Collins, and "Village Teacher" by Jack Sheffield.

English teacher Yakovlevna said that it was important to work out the first thirty pages of a book in detail, as they reveal the essence and content of the story, and the words that appear there are often repeated throughout the book. And of course, in the lessons, we developed elementary and communicative skills in speaking, reading and writing. At that time, a newspaper, "Moscow news", was published in English. We read and translated it with great interest. Today is different, as it is easier for young people to learn

Fig. 2.22   Working on EPR spectroscopy in Institute of Combustion Problems, Kazakh National University, 1973.

foreign languages, since there is a large amount of literature available, in addition to television programs and so on.

In 1979, the Ampere Scientific Congress on Magnetic Phenomena was held. I prepared for the oral report to be delivered there in the most thorough manner. Yakovlevna, seeing my zeal, offered to help me, so, for 3 months, we began to study at 7 a.m. and at 8 a.m. I continued to study in the class with all the "fellow students". Thanks to intensive training, my English language ability began to improve and I could easily speak it now. Even now, when I remember that time of intense learning, I think about her honesty, selflessness and dedication. For any class nowadays, the tutoring service is free. However, Yakovlevna always worked with pleasure and did not even think about any "compensation" from me. I did not think about this either. That was Tamara Yakovlevna! I am grateful to her.

In 1980, I passed a competition at the Ministry of Higher and Secondary Special Education of the USSR and went to England for a 10-month scientific internship. In 1981, I passed a competition for a scientific internship in the UK (United Kingdom) for 10 months under a joint program of the USSR Ministry of Higher and Secondary Special Education and the British Council. Such trips were very rare at that time. Each trip was a great event for my family. I remember how my whole family and father saw me off at the airport terminal in the center of Almaty. On the second floor of the airport terminal was the restaurant, where Tair organized a gala dinner at which my father advised me on successful scientific work.

My first long foreign scientific internship was in the early 1980s. I spent it in the UK. In Soviet times, the Ministry of Higher and Secondary Specialized Education of the USSR annually sent about 150 candidates of science in various fields for internships at the leading Western universities and research centers. The competition was held on the basis of exams in foreign languages, depending on the country of the internship, and on the basis of the results of scientific research, publications in scientific journals (including international ones).

The first four weeks of the trip to the UK were spent at the University of Essex located in the county of Essex in the city of Colchester. There, our training took place using audio and video materials. The first lessons were recorded, and we listened and watched them. Each trainee lived with an English family. All these activities contributed to the improvement of my spoken English. After this, for nine months, I worked at the Department of Chemical and Biochemical Technology (Professor Peter Roy, Department Head) under the guidance of Professor Anthony Burges) at University College London (UCL, London, UK) (Fig. 2.23). UCL is ranked from eighth to eighteenth in the four major world university rankings. During the internship, I studied cool hydrocarbon flames ("cool" meaning not very hot) stabilized on a flat burner using mass spectrometry, which is an experimental technique invented by Francis W. Aston (1877–1945, the winner of the 1922 Nobel Prize in Chemistry) for identifying molecules through the measurement of the masses of the molecules and their fragments. I also performed a very large review of the status of the study of the structure of hydrocarbon-like flames.

In addition to the scientific value, the internship was of great interest due to opportunity to practice English. During my internship, I significantly improved my English proficiency. This is partly because in Colchester,

Fig. 2.23   During scientific internship at University College London, September, 1981.

each Soviet trainee lived with an English family. Classes were held using audio, video equipment, and television. On January 18, 1982, before leaving England, I made a presentation entitled "Premixed Flame Composition Profiles", which aroused great interest from the audience (Fig. 2.24).

After the internship in England in 1982, I returned to Kazakhstan full of new and reformed ideas. I correlated all our university realities with what I saw in England. Two or three months afterward, I spoke passionately and convincingly in a meeting in the assembly hall of Kazakh State University (KazSU). In particular, I mentioned about the study of foreign languages in KazSU and posed the question of whether it is conceivable to master English using foreign language classes that meet only twice a week. I also mentioned about the student life. Our buffets are small. Similarly, I posed the question of whether it is conceivable for all the students to be served in these buffets during the break. In addition, I mentioned that, in England, there were vending machines. When a token had been inserted, the vending machine immediately gave a sandwich and a cup of coffee, with or without milk, as your heart desired. I asked if it was really so difficult for us to install such machines.

In 1987, I had the pleasure to attend the seminar of Professor Yakov Borisovich Zeldovich (1914–1987) (Fig. 2.25). He was a Soviet physicist and physicochemist, Academician of the Academy of Sciences of the USSR,

IMPERIAL COLLEGE

Department of Chemical Engineering and Chemical Technology

Prince Consort Road

London SW7 2BY

COMBUSTION RESEARCH COLLOQUIA

Spring Term 1982

▄▄▄▄▄▄▄

Monday 18 January   Dr Z.A. Mansurov *(University of Alma-Ata and University College London)*

"Composition profiles in premixed hydrocarbon flames"

Monday 1 February   Professor T. Takeno *(University of Tokyo and Imperial College)*

"Experimental studies on turbulent jet diffusion flames"

Monday 15 February   Dr W. Bartok *(Exxon Research & Engineering Company, Linden, New Jersey)*

"Studies on the chemistry of nitrogenous and sulphur species in combustion"

Colloquia to be held in Lecture Theatre One at 3.45 p.m.

Tea will be available from 3.30 p.m.

Visitors welcome without fee or formality.

Synopses of Colloquia will be available a week beforehand and will be sent on request. Please address any enquiries to :

Professor F.J. WEINBERG of this Department,

telephone number 01-589-5111, ext.1906.

Fig. 2.24   Announcement of my presentation in Imperial College, London, UK, in 1982 as a part of my internship there.

and Doctor of Physical and Mathematical Sciences. In addition, he was Thrice Hero of Socialist Labor, and Laureate of the Lenin Prize and four Stalin Prizes. He was known for his prolific contributions in cosmology and the physics of thermonuclear and hydrodynamic phenomena.

In 1994, I spent a month in the laboratory of Prof. H.G. Wagner (1928-) in University of Göttingen, Germany, under the funding of DAAD (German Academic Exchange Service), which is the world's largest funding organization for the international exchange of students and researchers. Wagner is

Fig. 2.25   With Yakov Borisovich Zeldovich in 1987.

a well-known scientist in the field of combustion and soot formation. Since 1971 he was director of the Physico-Chemical Institute of the University of Göttingen and Head of the Reaction Kinetics Department and Member of the Board of Directors of the Max Planck Institute for Flow Research in Göttingen. In 2015, I was pleased to present my book on "Soot Formation" to him in Budapest (Fig. 2.26).

Research requires funding to support personnel and purchase equipment and supplies. I am thankful for the numerous research grants that I have received over the years from Kazakhstan and other countries. For example, in 1995, 2003 and 2006, I received grants from DAAD. The funding is not only important for me, but are critically important for the research of my students. I consider teaching to be a very important part of my work. The courses that I have given include "Physical Methods of Research" and "Chemical Physics" in Kazakh and Russian languages, and "Soot formation", "Basis of nanoscience and nanotechnologies" and "Self-propagating high-temperature synthesis" in the English language.

Chemical physics and large-scale studies of combustion processes is a fairly young industry in Kazakhstan. We relied on the tremendous support and authority of leading scientists in Kazakhstan and Russia. In addition to what I mentioned above, I would like to note the role of Alexander G.

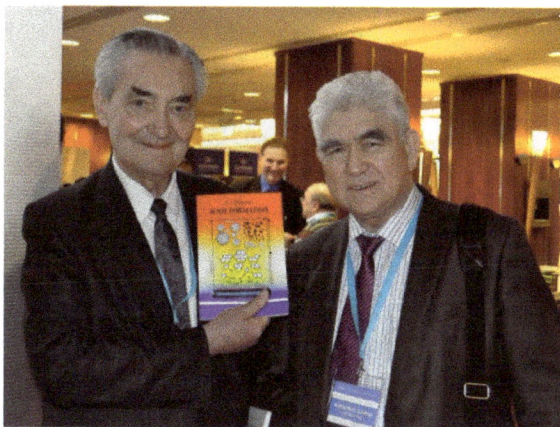

Fig. 2.26   Presenting my book on "Soot Formation" to Professor H.G. Wagner in Budapest in 2015.

Merzhanov (1931–2013) (Fig. 2.27), who developed combustion theory and self-propagating high-temperature synthesis (SHS) technology. He is the Director of Institute of Structural Macrokinetics and Materials Science (ISMAN), Russian Academy of Sciences, Chernogolovka, Moscow, Russia. At Al-Farabi Kazakh National University, he facilitated the formation of the Institute of Combustion Problems (ICP), where I have been working.

The kinetics of a chemical reaction refer to the rate of the reaction. This rate is important for the industrial implementation of any reaction. In 1990, I defended my second doctoral dissertation on the topic "Nonisothermal Cool Flames of Hydrocarbons" at the Institute of Structural Macrokinetics of the USSR Academy of Sciences in Chernogolovka, Moscow, Russia (Fig. 2.28). The research of this Institute focuses on the macroscopic kinetics of chemical reactions. Merzhanov gave me significant support. He appointed Vilen Vagarshevich Azatyan, an outstanding scientist in the field of chain reactions, as my Advisor. Four months after a detailed study of the material and heated discussion at the Institute's seminar, the work was submitted to the Dissertation Council. In response to the comments made, I did not agree with all the comments and began to give explanations. Then G.I. Ksandopulo, who was presented at the defense, whispered to me, "Zuklhair, at least you agree with some comments." Summing up the defense, A.G.

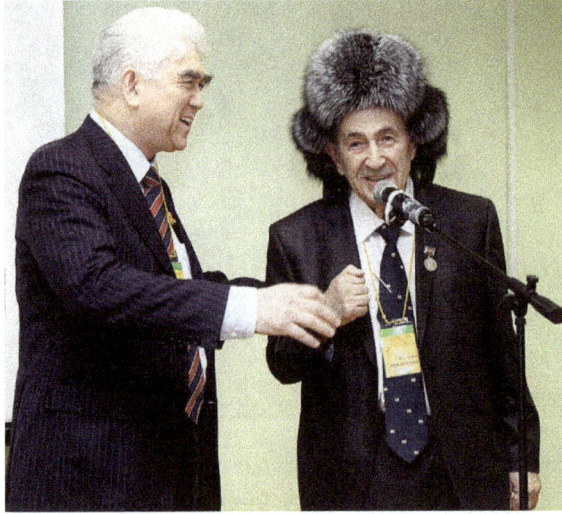

Fig. 2.27   With A.G. Merzhanov on his 80th birthday, Chernogolovka, Moscow, Russia, 2011.

Fig. 2.28   Institute of Structural Macrokinetics of the USSR Academy of Sciences in Chernogolovka, Moscow, Russia. (http://www.ism.ac.ru/n_about/)

Merzhanov noted that a large amount of experimental work had been completed, and interesting new data that are good material for theorists had been obtained. Thus, I received a Doctor of Science degree from Institute of Structural Macrokinetics, Moscow, in 1990. This is my second doctorate degree.

Heavy crude oil is highly-viscous oil that cannot easily flow in an oil well. Its density is higher than that of light crude oil. On the other hand, oil sands are loose sands or partially consolidated sandstone that contains a dense and viscous form of petroleum known as bitumen. In 2015, I gave a keynote lecture on "Processing of heavy oils and oil sands" at the 3rd World Congress on Petrochemistry and Chemical Engineering held in Atlanta, Georgia, U.S.A. (Fig. 2.29). Over the years, I have lectured in countries including the United Kingdom, Germany, Italy, Poland, Turkey, China, Belgium, U.S.A., Taiwan and France.

2.2.5 *Sports in my life*

Scientific research performance requires depth of scientific interests and passion for the research. However, the physical condition is important.

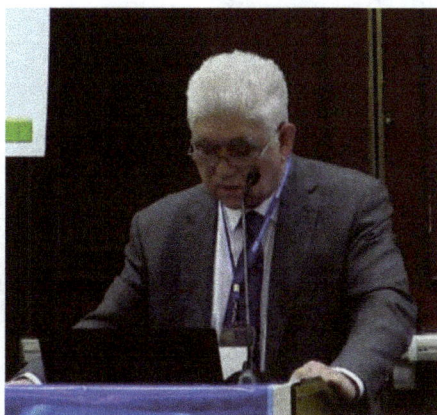

Fig. 2.29 Speaking at the 3rd World Congress on Petrochemistry and Chemical Engineering in Atlanta, Georgia, U.S.A., 2015. (https://petrochemistry.conferenceseries.com/speaker-details.php?confyear=2015&speaker=zulkhair-mansurov-institute-of-combustion-problems-kazakhstan)

Sports have always helped me in this regard. In fact, sports constitute a special page of my life.

Even before entering the university, I was engaged in various types of sports. In seventh grade, I was engaged in classical wrestling in Panfilov. While I studied in the First High School in Talgar, I played basketball under the guidance of Shora Kadyrovich Yumashev, who was the USSR Honored Basketball Coach, and the State basketball coach of the Republic of Kazakhstan, in addition to the Secretary General and Vice President of the Basketball Federation of the Republic of Kazakhstan. Talgar is famous for its mountains (Fig. 2.30). In 2004, I visited Tuyuk-su Glacier in Almaty. There, I had a great time with my grandson Nur-Kadyr (Fig. 2.31).

During my second year in the university (April 1965), I participated in a High Jump competition in a sports festival (Fig. 2.32). In 1968, I won the city championship among students by jumping over a distance of 1.85 meters. After that I was accepted into the student team of the Republic and we went to the All-Union competition in Kiev, where I repeated my high jump result and reached my best result in triple jump (13.82 meters) and the

Fig. 2.30 Pik Talgar (a northern 4979-meter peak in the Tian Shan mountain range) in the Talgar moutains. (https://en.wikipedia.org/wiki/Pik_Talgar, public domain)

Fig. 2.31   Visiting Tuyuk-su Glacier (3450 m) with my grandson Nur-Kadyr, 2004.

hammer throw (54 meters). For an amateur athlete, this was a pretty good result.

As a student for five years, I was active in sports. During the year in which I served in the Soviet Army near Panfilov, I climbed very day, followed by running for 3 km to the mountains and then back in boots and with a bare torso. In my everyday crush and constant labor, I tended to ignore many of the deepest, most important values of life.

In 2010, a table tennis tournament was held in memory of my late wife, Raushan Magzumovna, who was a master of table tennis (Fig. 2.33). I cherish the support of numerous friends.

2.2.6 *My scientific work*

I have established a scientific school in the field of chemical physics of combustion processes and nanotechnologies. Under my supervision, more than 50 Ph.D. students have successfully defended their dissertations. They include three specialists from Egypt, one from China and one from Cuba. Among my graduates are a Dean of the Chemistry Department and Heads

Fig. 2.32 Participating in the High Jump student competition in a sports festival in the university, 1965.

Fig. 2.33 Table tennis tournament dedicated to the memory of my wife, Raushan Magzumovna, 2010.

of various laboratories in universities and research laboratories all over the World.

My research field mainly concerns the study of the rate and mechanism of the combustion of hydrocarbons. A cool flame is a flame that has a maximal temperature less than about 400°C. A sooty flame is a flame that involves incomplete combustion, so that carbon particles remain. My research also pertains to the structure of cool and sooty flames. In addition, my research relates to the synthesis and investigation of nanocarbon materials for various functional uses. In my work, hydrogen atoms were discovered in the cool flames of hydrocarbons.

Carbon nanomaterials (e.g., carbon nanotubes and graphene) occupy a leading position in modern materials science. They can be obtained by the combustion of agricultural products, such as rice husk, walnut shell, apricot stones etc. Practical uses relate to environmental, medical, agricultural and other applications. Kazakhstan scientists have been successfully working in the synthesis, behavior and applications of carbon nanomaterials. For 20 years, my co-workers and I have conducted research on nanomaterials obtained by burning agricultural waste products. On the basis of the research results on the synthesis of fullerenes, carbon nanotubes, special (superhydrophobic) soot and graphene in the flame, it is possible to use a scheme (as proposed by Henning Bockhorn, a specialist in soot formation in combustion, Karlsruhe Institute of Technology, Germany) that allows the formation of fullerenes at low pressures and soot at high pressures. In addition, the scheme involved the formation of graphene as an intermediate product in the process of soot formation. Moreover, a scheme for soot formation has been developed for any type of hydrocarbon fuel.

Since 2003, with my direct participation, a large-scale project on "Electronic Portals of Universities of Central Asia" has been carried out with Robert Schumann University (Strasbourg, France). The objectives of the project are to use new information technologies to create the foundation of an electronic university and to introduce the European experience in building basic e-learning structures to the universities in Kazakhstan, Uzbekistan and Tajikistan.

The publication of peer-reviewed research papers is important for the dissemination of new knowledge. Journals provide an important archival avenue for this purpose. A separate part of my scientific activity concerns the founding of the English-language journal "Eurasian Chemico-Technological Journal". The first issue was published in December 1999. In response to the first issue, N.A. Nazarbayev (1940-), the President of the Republic of Kazakhstan, wrote, "Dear members of the scientific community! Congratulations on the release of the new Eurasian Chemical-Technological Journal! I sincerely hope that the new edition will play a significant role in the development and strengthening of scientific ties between scientists from Asia, Europe and the entire world community."

In 1992, I was appointed vice-rector for scientific work of the Kazakh State University (currently known as Al-Farabi Kazakh National University). In this capacity, I was involved in the creation of several scientific and research institutes, a scientific and technological park, scientific centers, and the organization of exhibitions of scientific achievements. In 2000–2019, I served as the General Director of the Institute of Combustion Problems. In 2001–2010, I served as the first vice-rector of the al-Farabi Kazakh National University. In this capacity, I attended in 2008 a meeting in Washington, D.C., U.S.A., for creating a global educational network. In this meeting, I met U.S. Secretary of State Condoleezza Rice (1954-) (Fig. 2.34).

Fig. 2.34 With U.S. Secretary of State Condoleezza Rice, The White House, Washington, D.C., 2008.

2.2.7 *My advice to young people*

To be successful in science or any other profession, one must have a purpose in life. This means having a meaningful goal, with recognition that this profession is your assignment in life. With this notion deep inside one's heart, one can persevere and strive in the midst of adversities. However, one must also recognize the value of the family. The love among family members is beautiful and must be cherished. The love of one's country is also important.

Scientific advancement requires creativity, which largely determines the impact of the scientific work. However, in the initial stage of the formation of a scientist, the rules of scientific ethics and the issues of interaction between science and society need to be understood. Young scientists must correctly understand what can be done and what cannot.

Reading is important. Read not only scientific literature, which is certainly necessary in view of professional activity, but also fiction, and even science fiction. In the modern world, it is not enough for a scientist to be just a good experimentalist. It is necessary to state correctly their thoughts and opinions in a written form. Science fiction helps to expand the vocabulary and realm of thought. Moreover, it allows the inquisitive minds of young scientists to go beyond the scope of conservative thinking, and allows one to take a fresh look at the object of study, and may even push one to new discoveries! Specifically, I have the following advices to offer.

Firstly, the student should be aware of the history of his/her educational institute, the outstanding scientists who worked and work in it, and its leaders. This would help the education and promote aspirations for achievement.

Secondly, intensively study foreign languages. Today, without knowledge of Western languages, especially English, it is very difficult to get a job or to do science. Careful attention should be given to the choice of the discipline of the study and the possible specialization within this discipline. This choice will greatly affect your career path.

Thirdly, you should have the skills to work with information and computer technologies at the level of an experienced user. In today's world, where information technology plays a leading role in the development of society, it is simply ridiculous not to be able to work on a computer. Do not be restricted in where you study. Visiting other educational institutes, making friends in various places, and developing a broad mind in relation to life and career would help.

Fourthly, starting from the first year of study at a university, a student should be involved in scientific activities, participate in ongoing scientific events, competitions, conferences, etc., and read broadly. Informal discussions provide a good opportunity to put your scientific knowledge to the test in the process of scientific discourse, particularly in relation to controversies. Do not hide your doubts inside yourselves, but strive to look for the right approaches and methods to solve a scientific problem. Also try to be innovative and do not exclude the possibility of starting a technology business.

Lastly, be sure to do sports. This is not only good for your health, but also enhances your performance and adds to your enjoyment in life.

Chapter 3

Merton C. Flemings – from America to the world through metal solidification science

3.1 Introduction by the Editor

The solidification of liquid metal is one of the most common ways to form a metal solid. By using a mold with chosen dimensions, the shape and size of the metal solid can be controlled. This process is involved in casting, which is a key method for forming various shapes of metal articles.

The metal solidification process is also important for welding and soldering, which are important for joining. Welding is used to make complex structures, such as pipe manifolds. Soldering is widely used in the electronic industry for electrical connections. The reliability of solder joints is critical to the reliability of microelectronics such as computers. In addition, the metal solidification process is commonly used in three-dimensional (3D) metal printing and the fabrication of metals that are filled with ceramic particles. Such metal-ceramic composites are used for cutting tools and electronic enclosures, due to their attractive mechanical and thermal behavior.

Due to the high temperatures of the liquid metal in the solidification process, the control of the process for obtaining a nonporous solid with the desired structure in the microscopic level (known as the microstructure) is non-trivial. Aspects of the microstructure include the size, shape and orientation of the grains, in addition to various types of microscopic defects. The microstructure largely governs the mechanical behavior (e.g., the strength) of the resulting solid metal. By controlling the conditions of the solidification process, the microstructure can be controlled, and in turn the behavior of the resulting material can be controlled. For this purpose, detailed scientific understanding of the solidification process in relation to the heat flow and mass flow is required.

The most respected pioneer in the world in the field of metal solidification science is Professor Flemings of the Massachusetts Institute of Technology (MIT). The science that he developed and unraveled has provided the foundation for the metal industry.

Professor Flemings contributed to the shaping of materials education and the training of a generation of materials scientists and engineers (including myself). In MIT, he is Toyota Professor Emeritus and the director of the Lemelson-MIT program, which is aimed at honoring inventors and inspiring inventiveness in young people.

3.2 Life experience as told by Professor Flemings

3.2.1 *Early years*

I come from a long line of New England farmers, merchants, and Protestant ministers. The side of the family I knew best, in terms of both its history and its impact on my life, descended through the male lineage from John Dexter, who served in the army during the American Revolution beginning in 1775. After his term in the army, he moved to Pomfret, Vermont, cleared land for a farm, married and started a family. His youngest son, Parker Dexter, married Betsy King in 1820. They subsequently purchased 150 acres of unbroken wilderness in West Topsham, Vermont to start their own farm. West Topsham is a part of the township Topsham, which at the time had only about 1,000 residents. The farm was to remain in our family through the Second World War, eventually comprising 1,000 acres. Fig. 3.1 shows the farm as I remember it in my youth. Parker and Betsy remained at the farm their entire lives, siring 15 children, 12 of whom survived to marry and build their own lives and families. Fig. 3.2 shows the couple at their 50th wedding anniversary. Today, much of the farming in Vermont is a thing of the past, but the state remains a beautiful and restful place (Fig. 3.3).

The Dexter children, as they reached maturity, moved on to build their own lives in farms, business, and the ministry. Among them, Solomon King Dexter was an entrepreneur. After engaging in several businesses, including

Fig. 3.1   The family farm in West Topsham, Vermont as it was in the late 1930's. At the far left are a small house in which we lived when we visited, as well as a chicken coop. Moving to the right is the main home, and attached to the home is the horse barn with a tool shed at its right. Farther to the right is the large cow barn that also housed the hay, and to the far right is a pig sty.

Fig. 3.2   My great-great grandparents, Betsy and Parker Dexter in 1870, on their 50th wedding anniversary.

supplying goods to Civil War troops, he started what became a successful provision and grocery business in Lowell, Massachusetts, while inheriting and managing the farm in Vermont.

Operation of the farm and store later fell to Solomon's son, my grandfather Royal King Dexter. By the beginning of the Second World War, the store included a semi-automated factory canning "Dexter's Best Baked

Fig. 3.3   A typical view of rural Vermont showing a covered bridge and rolling hills.

Fig. 3.4   My mother and her six siblings, about 1920. My mother, Marion Dexter, is the oldest, on the far left.

Beans." I was, then ten years old, fascinated by the machinery, as well as by the fact that the process still required a man manually dropping a small piece of salt pork into each can just before the machinery pressed the cover on.

Royal King Dexter had six children, my mother Marion included, who are shown in Fig. 3.4 in the early 1920s. My mother married my father, Merton Flemings, in 1925. Both had spent their young lives in Lowell,

Fig. 3.5 My family, Worcester, Massachusetts, 1940, showing my mother and father, sister Eleanor, baby John, and me.

but were soon to move to Syracuse, New York. My father's position there was as a salesman for Carter Rice Company in upstate New York selling paper to printing establishments. I was born in 1929, just days before the stock market crash and the beginning of the country's Depression years. My father was fortunate to remain employed by the same company through the Depression, and I had the pleasure of hearing about many far away sounding places like "Canajoharie" and "Chittenango" (towns in upstate New York).

When I was in the third grade, my father was promoted to be the head of the Worcester, Massachusetts office of his paper company, and so I lived there until I went to college in 1947. Worcester was then an important and prosperous manufacturing town producing heavy machinery, metals, and ceramics, with an engineering school, Worcester Polytechnic Institute (WPI), not far from our home. Fig. 3.5 is a family photograph taken when I was about 12 years old.

My recollection of my early elementary school days has dimmed with time, but two events stand out: the hurricane of 1938, and Pearl Harbor Day, 1941. I remember the hurricane as a day of wind and excitement, and Pearl Harbor as a life changing day. During the war I collected a great deal of newspaper from nearby homes to bundle for recycling, aiding in the war

Fig. 3.6   A buggy ride in Vermont, 1945, three cousins and me.

effort and, incidentally, providing me a small amount of cash. I had a small workshop in the basement of our house and was a devoted builder of model airplanes, especially the fighter planes of all combatants. I was conscious of the rationing of gas and certain foods. On the other hand, we were not hungry, and life for a young boy growing up was on the whole better for my family than for many Americans during the war years.

We went with some regularity in summers during the pre-war and war years to "the farm" in Vermont. Fig. 3.6 shows three cousins and me taking a buggy ride in 1941. In the summer of 1945 when I was 15, I went and stayed in the home of an aunt who lived near the old farm, and provided her some help while her husband was in the war. In addition to the help I gave her, I worked part time with the new owner of what had been our old family farm, Fig. 3.1. My main job was driving a horse drawn raker and helping stow the hay. I was privileged that year, and in my earlier years in Vermont, to have seen a world that no longer exists. It was one where electricity was still coming to many homes, where horses were still a widely used source of power, and where mechanical devices, powered and unpowered, were still simple enough to be fully understood by a young boy.

I was always an avid reader and also consumer of news, through newspaper, radio, and magazines. My favorites were Time and Life magazines and Popular Mechanics. I was already aware of how rapidly the world was changing technologically, with the advent of jet aircraft, radar, etc. That awareness was intensified by the dramatic news in mid-August 1945 of the invention and dropping of the atomic bomb. I was not wholly surprised, having remembered an article in Popular Mechanics some years earlier that had a sketch of how "atomic energy" could someday power an automobile. This article was certainly published before a wartime edict was established banning all mention of atomic energy.

Overall, I remember my school years in Worcester as being a suitable, but not especially exciting, time of preparation for a future that I was determined to make successful. In addition to engaging in the summer activities referred to above, our family hiked a lot, with the result that I gained a lifelong love of hiking anywhere, but especially in the White Mountains of New England. I won a spelling bee of citywide significance when I was 11, surely a result of my avid reading.

Two high school teachers stand out in my memory as having had an impact on my life's trajectory. One was the woman who taught me geometry. I do not know how they teach it now, but she taught it the way Euclid did. . .step by step, building the argument one irrefutable step at a time. It was the first time I saw beauty in mathematics. The other important teacher was my physics teacher who told me (and my parents) that even though my grade in his class would be an "A," I was not living up to my potential and should put more effort into the subject. So I did, not to please him or my parents, but I now think because of the confidence his remark instilled in me.

My college application process was completed in early 1947 and I was accepted at Cornell, MIT, and WPI. I had applied to Cornell as an English major with the possible aim of following in the footsteps of an uncle who had been a wartime correspondent and was killed in the early stages of World War II. Also, I enjoyed reading and writing. However, engineering was even more attractive to me. My parents would have much preferred (for financial reasons) that I stay home and enroll at WPI, but I wanted to strike out on my own. I also understood MIT was a very demanding school and I was looking for a tough challenge. I feel everlasting gratitude to my grandfather who encouraged me to go to MIT, and to my parents for acceding to my wish to do so. I often feel that my life really began the moment I entered the halls of MIT, Fig. 3.7.

Fig. 3.7   The "Great Court" of MIT.

3.2.2 *College years*

Once on MIT turf, I never looked back, partly because of the excitement of so much that was new, partly because of the challenges of each day. One feature of my freshman year unique to that class was the fact that 1947 was the peak year for college enrollment of veterans returning from WWII. My class at MIT was about 50% veterans. The half of us who were fresh out of high school were in awe of our older classmates who seemed much better equipped than we to deal with the rigors of MIT. I am grateful for the lessons I learned from them.

Another important (for me) feature of my freshman year and of my entire undergraduate years was the fraternity I joined. Fraternities and sororities in recent years at many colleges have earned a deservedly poor reputation, but I want to describe how mine was different. My fraternity, like the school as a whole, included many veterans, all displaying a maturity gained from their wartime experiences. Their primary interest was obtaining an education and moving on to a new life. They were good role models. The fraternity had a code of conduct including study rules that were rigorously enforced by the older brothers. Meals were a dignified affair. Wearing a jacket and tie was required. Everyone waited for the house president to signal the beginning of dinner and did not leave the table until he struck the gong. Several of the older brothers became particularly important mentors.

Many of them remained longtime friends. I gained both confidence and a higher level of aspiration from these individuals.

A famous comment about an MIT education is that it is like taking a drink from a firehose. That, I would say, was my experience in my first months at MIT. Nonetheless, as the end of the first term approached (then in early February), I realized that I had made the grade and that MIT would be my home for the full four years.

After the end of my freshman year, I joined two high school friends to take jobs in the U.S. Forest Service in McCall, Idaho, work I repeated the following summer. They were wonderful summers, being outdoors, with good physical activity, and a respite from lectures and labs. In the first part of each summer, we "cleared brush," that is, chopped up the tops of the trees left by the loggers, and piled the branches for later burning. In both Augusts, however, we were full time fire fighters, a job which had its own exciting moments. It also paid well, with much overtime at 50% over our base wage of $1.25 an hour. We were on one such fire in central Idaho on August 5, 1949 when we received word of the deaths of 12 Forest Service "Smokejumpers." These men have been memorialized in Norman Maclean's book Young Men and Fire.

By mid-term of my sophomore year, it was time to choose a specialty within engineering. I was attracted to both mechanical and chemical engineering, but finally settled on metallurgy. Metallurgy seemed to me to combine many aspects of those two fields and was a field where jobs were plentiful. Perhaps my decision was also influenced by my respect for my mother's cousin who had graduated in metallurgy from MIT and then became a successful patent attorney. But probably the strongest influences came from two of my fraternity brothers, both veterans, who were enrolled in the department. One of them, 95 years old at this writing, is as ebullient and optimistic now as he was then.

My four undergraduate years were mostly happy and were certainly memorable. We students felt proud of our accomplishments and as the senior year wore on we became focused on next steps. A number of us graduating in 1951 decided to stay on and were admitted for graduate study at MIT. Among them were myself and two friends in the class. We three chose the same faculty member to be our thesis supervisor and decided to share an apartment together. We were to spend all our graduate years living and working happily together, receiving our doctorates at slightly different

Fig. 3.8 MIT graduation, 1951. My roommates through my graduate student years and I are the three on the left: from left to right, Edward Hucke, me, and David Ragone, and on the right is another good friend, Richard Willard, who also roomed with us for much of those years.

times in 1953. What little competition there might have been among us was overshadowed by the companionship and collaborations that evolved. Fig. 3.8 shows the three of us at our Bachelor degree graduation, along with another good friend (on the right) who lived with us for most of our graduate years.

Two central elements of our graduate engineering education were the oral exam and the thesis, a study expected to be an original technical contribution. I passed my oral exam in spite of experiencing marked stomach pains during its progress. I had my appendix removed later that afternoon and was grateful that one professor stopped by my bed the next day to enquire after my health. My roommates and I had the enormous advantage of having a thesis advisor who gave support and suggestions but left the research details and even its goals to each of us. At one point in early July 1953, he decided we needed a break from our laboratory work and "ordered" that we take the weekend off. We agreed and spent three days at what for us was an expensive hotel, just across the bay from where Jack Kennedy and Jackie Bouvier were celebrating their engagement (which had been announced in late June). We returned to MIT much refreshed and, I believe, much more effective in our research during the remaining months. My two

Fig. 3.9 Last days of thesis experimental work in September, 1953. To speed our progress and have unimpeded use of needed equipment, we often worked through the night. Here I am taking a break on my 24th birthday. Co-experimenter Edward Hucke is in the background.

roommates and I all finished our graduate work before the end of 1953. Two of us had experiments progressing that often required all night in the laboratory (Fig. 3.9). After graduation, we went our separate ways, one of us to a teaching job and two of us to industry, but have kept in touch ever since.

3.2.3 *Two years in industry*

I chose the industry path and went to a company then called American Brakeshoe Company. My job in a foundry laboratory there was to apply concepts I had learned in my graduate work on the solidification of metals and, in particular, how the properties of aluminum castings could be improved through new methods of solidification control. With some trial and error, we developed practical ways of doing this, and by the end of my two years there, we were shipping this new type of "Premium Quality" aluminum castings to several sources, including Grumman Aircraft Corporation. The process we developed provides the basis of government and industry specifications today and is in wide industrial use for aircraft and automotive applications.

Satisfying as those two years in industry were, I missed the academic life and began to think of applying for a teaching position at a research university. This was a time of huge expansion of engineering schools in the United States, with many job openings. I used my two-week vacation in the summer of 1955 to make a circular trip visiting universities, starting clockwise at Penn State University, then Iowa, Michigan, Cornell, and finally MIT. Several of the universities indicated strong interest. One of them offered me a position only a few weeks later as Associate Professor in charge of their foundry laboratory. I was tempted, but when I was later offered a position at MIT as Assistant Professor, I did not hesitate to accept it.

3.2.4 *Teaching, Research and Administration at MIT*

I began my teaching and research career at MIT in early 1956. On the research side, I continued engineering studies on metal casting processes. The foundry we had at that time was quite large, with industrial scale melting and molding equipment, including the large melting furnace shown in Fig. 3.10. Working with a local artist, we developed a molding process to enable artists to make statuary at much lower cost, a process that has been widely used since that time for manufacture of automotive and marine engines. Fig. 3.11 shows one such casting emerging from its mold. I wrote a book during that time jointly with a senior faculty member on foundry practice that is still used in some parts of the world today.

As the years went on and as teaching at MIT evolved to that of greater emphasis on engineering science, my own research turned more strongly toward the basic engineering science of casting, particularly to solidification processes. When the foundry became my full responsibility in the early 1960s, the industrial scale equipment became irrelevant to directions I wanted our research to take. Instead, we went to small scale experiments and computation, Fig. 3.12 shows a student, Kenneth Young, conducting an experiment on semi-solid metal flow behavior. Fig. 3.13 shows a student, Qi Zhao (with the MIT sweatshirt), stretching after a long session of computation. Fig. 3.14 shows a student, Andreas Mortensen, engaged in structural studies. Andreas later rose to be a full professor at MIT, later Professor and then Vice President for Research at EPFL (Swiss Federal Institute of Technology) in Lausanne, Switzerland.

Fig. 3.10   Melting and pouring a heat of cast iron in the foundry laboratory, 1956.

Fig. 3.11   A casting made by the new method we developed, now called "Lost Foam" casting. The pattern for the casting is made of a foamed polymer that burns out as the metal fills the sand mold.

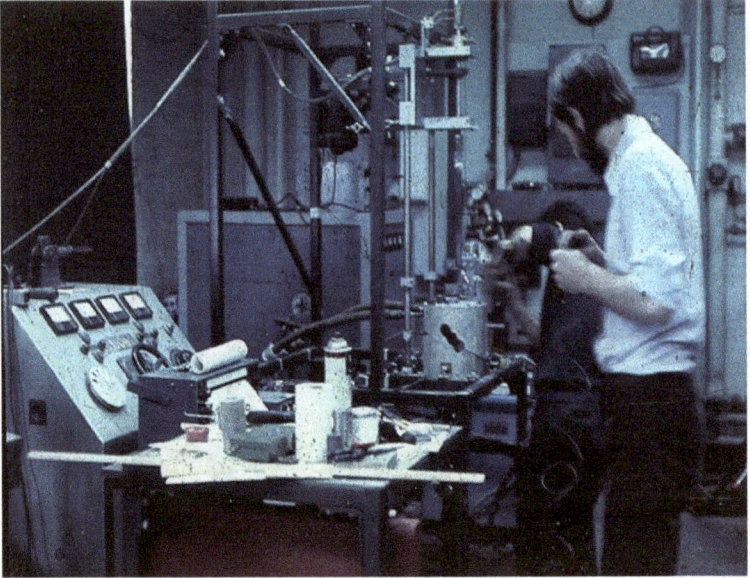

Fig. 3.12   A graduate student, Kenneth Young, conducting an experiment on forming semi-solid metals.

Fig. 3.13   A graduate student, Qi Zhao, in the course of a long period of computer modelling of experiments.

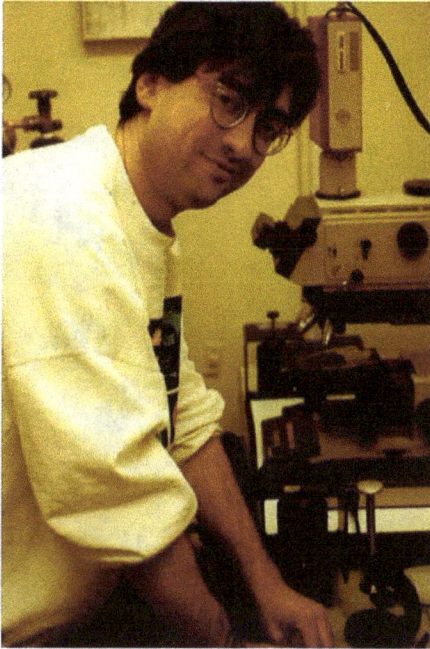

Fig. 3.14 A graduate student and later fellow MIT faculty member, Andreas Mortensen, conducting a metal structure analysis.

It was exhilarating in these studies to be able to dig deeply into the engineering science of metal solidification–to ask and to answer in new ways questions such as "Why is the centerline of continuous castings often richer in alloy element than the outer portion?" and "Why does the internal structure of a solidifying alloy depend on cooling rate?" It was also gratifying to be able then to use the answers to such questions in the development of new and improved processes and products. Mostly, it was wonderful to work with bright young students in developing together both the answers to the questions and the search for practical applications. I loved to teach and wrote a book Solidification Processing based on lecture notes from a graduate course I taught during this time. The book, Fig. 3.15, was published in 1974 and can be found still today on the desks of metal technologists throughout the world.

The book and of course our research accomplishments owe much to the graduate students and visiting researchers in my group, many of whom

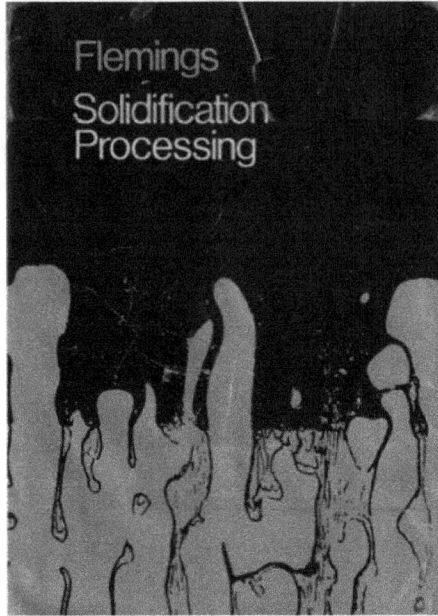

Fig. 3.15 Cover of the book "Solidification Processing".

came from outside the United States. It is a joy for me to be able to remain in contact with them and to follow their successes in industry and academia throughout the world.

By the late 1970s, I was looking for a new challenge, and was given an opportunity to lead a program then being negotiated whereby MIT would build a two-year technical college in Shiraz, Iran. We undertook the task jointly with Wentworth Institute in Boston and a Cambridge architectural firm. On the Iranian side, the project was sponsored by a foundation directed by Princess Ashraf, the sister of the Shah. The project was fully successful, and by the end of 1978, we had the building built and the first class of students approaching graduation. Sadly, certainly for the students involved, the school was closed at the time of the deposition of the Shah in February 1978. My trips to Iran were a pleasant and enriching experience, especially when my wife Elizabeth could accompany me, as seen in Fig. 3.16, a photograph from a sightseeing weekend.

There was a growing sense in the Engineering School of MIT at this time, shared by me, that MIT in its research and teaching should place more

Fig. 3.16   My wife Elizabeth and our hosts on a weekend sightseeing trip in Iran.

emphasis on societal and industrial needs in its teaching and research. I sought and was given the opportunity in 1979 to build an interdepartmental center, a "Materials Processing Center," that would seek to strengthen collaborations both within and outside MIT and would encourage industry relevant materials research. The Center soon had a substantial research budget and participating faculty members from across the School of Engineering. In the succeeding years, it grew to become one of the largest centers at MIT and today exists with a new name, the "Materials Research Laboratory," having joined forces with another laboratory.

In 1982, I became head of what had by now changed its name from Metallurgy Department to Department of Materials Science and Engineering (DMSE), a position I held for 12 years. Those years provided a different kind of exciting challenge. When I took over as head of the department, we were still a department whose faculty focused almost exclusively on metals. But the research opportunities and the jobs for graduating students lay across the materials field broadly. . . in plastics, ceramics, and electronic materials as well as metals. The goal was clear: We needed to develop a teaching program that dealt broadly with fundamentals applicable to all engineering materials and to conduct research over the full range of such materials. We

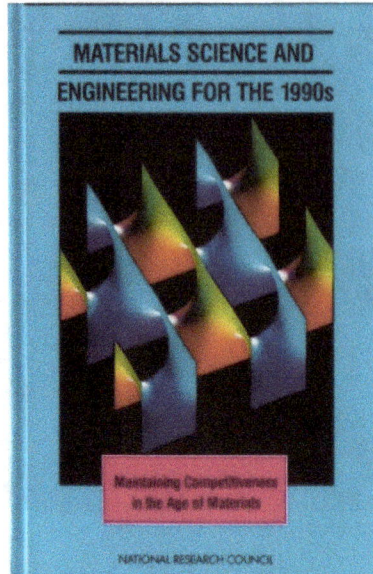

Fig. 3.17   The report of our national study of needs and opportunities in Materials Science and Engineering, 1989.

were proud of knowing we had mostly accomplished those goals by the time I completed my tenure as department head in 1995. Today, the interests of the department faculty members stretch still further than we had envisioned then to include biomaterials as well.

During these "Department Head" years, I spent time also on the national and international materials scene, encouraging other metallurgical departments to broaden their research and teaching. I co-chaired a National Research Council study over the last several years of the 1980s. We issued a report published in 1989 as "Materials Science and Engineering for the 1990s" (Fig. 3.17).

I was on sabbatical leave in Japan for much of the academic year 1989, joined by my wife and two young daughters. I was appointed Nippon Steel Endowed Professor at the University of Tokyo and divided my time between the University and the Nippon Steel Research Laboratory.

My wife, who was to become professor of Asian/Japanese art history at Boston University, pursued her professional interests in Tokyo, and our two children, aged 7 and 9 years old, attended Japanese public school. We

Fig. 3.18   Our daughters Elspeth (left) and Cecily (right) on one of our trips to Japan.

returned to Japan with our daughters in subsequent years on shorter trips that were always memorable (Fig. 3.18).

My main host in Japan was Professor Tasuku Fuwa, a long time friend, academic leader, and then a lead scientist for Nippon Steel. He is shown on the left in Fig. 3.19. On the right is then Nippon Steel researcher, Toshihiko Koseki, who later received his doctorate at MIT under my supervision. Dr. Koseki went on to have a distinguished career, first at Nippon Steel, then at the University of Tokyo as Professor and Vice President.

In 1997, I led a task force of 25 MIT faculty members of the School of Engineering to review the engineering programs of two universities in Singapore. We wrote and submitted our report in 1998. For the following two years, I co-chaired with a Singapore faculty member the first large collaboration of MIT with Singapore: the "Singapore MIT Alliance." It was a wonderful experience, working with a quite different culture in developing a distance collaborative research and teaching program.

My last administrative role at MIT was to lead for several years the "Lemelson-MIT Program," a program aimed to support and encourage the inventive process. In the course of my eight years in that role we started a program to encourage inventiveness in high school students and conducted

Fig. 3.19   Springtime during our Japan sabbatical, 1989. Left to right, Tasuku Fuwa, me, and Toshihiko Koseki.

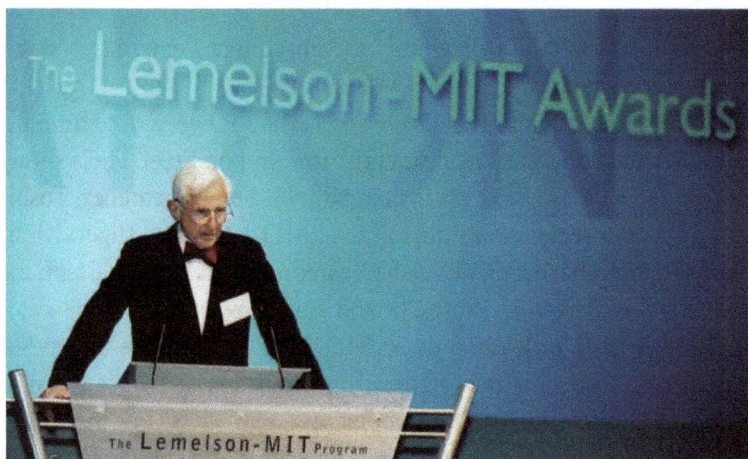

Fig. 3.20   Public presentation of the NSF supported report on invention, 2004.

a national study, partially supported by the National Science Foundation that was published as Invention, Enhancing Inventiveness for Quality of Life, Competitiveness and Sustainability. Fig. 3.20 shows me presenting the report at the National Academy of Engineering in Washington D.C. in 2004.

3.2.5 *Awards and honors*

I am grateful for having been the recipient of a long list of awards and honors during the course of my career. I was elected to the National Academy of Engineering in 1976 and to the American Academy of Arts and Sciences in 1980. I list below several awards from later years:

I was awarded the ASM (American Society of Metals) and TMS (The Materials Society) Distinguished Lectureship in Materials and Society for the year 2000, for the *"invention of numerous new solidification technologies which are widely used industrially, and for leadership in defining the national agenda in materials science and engineering."* I received an honorary doctorate from the Swiss Federal Institute of Technology in Lausanne in 2004 for my *"role as pioneer and for exceptional scientific contributions in the field of solidification and foundry."* In 2005, I received the Gold Medal of the Japan Institute of Metals. I also received in 2005 the Albert Easton White Distinguished Teacher Award from ASM *"in recognition of unusually long and devoted service in teaching as well as significant accomplishments in materials science and engineering and an exceptional ability to inspire and impart enthusiasm to students."* I was elected Honorary Member of AIME (American Institute of Mining and Metallurgical Engineers) in 2006 for my *"pioneering work in solidification processing, for the development of novel processes which are used commercially, for leadership in expanding the field of metallurgy to materials engineering, and to materials science and engineering, and for leadership in establishing a national agenda for the field of materials."* In 2007, I was awarded the Benjamin Franklin Medal of the Franklin Institute *"for outstanding contributions to understanding the fundamental and technological aspects of the solidification of metallic alloy."* Fig. 3.21 is a photograph taken at that event. Towards the end of my active days at MIT, I was honored by the establishment of an assistant professorship in my name. My name was also given to a refurbished foundry for student use.

3.2.6 *Family*

I have been blessed in my life by having a close and loving family, now including four children, two of them from a previous marriage. Fig. 3.22 shows three of the children who joined my wife and me walking in the Swiss

Fig. 3.21   Elizabeth and me at the ceremony where I received the Benjamin Franklin Medal, 2007.

Fig. 3.22   A family trip to the Swiss Alps, 1984. Left to right in the front, Peter, Cecily, Elspeth; in the rear, me and Elizabeth.

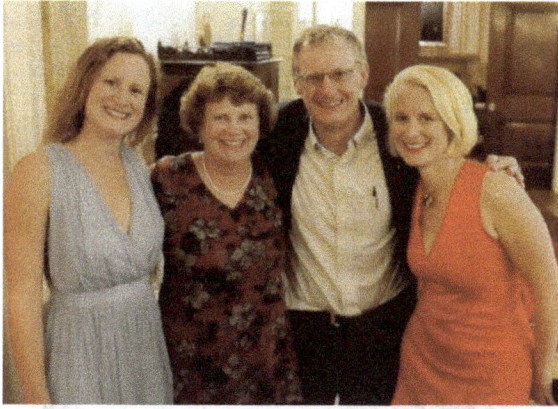

Fig. 3.23   A recent photo of all four children: left to right, Elspeth, Anne, Peter, Cecily.

Alps in 1984; Fig. 3.23 shows all four children 35 years later, in 2019. All are grown, married, and parents, blessing us with a total of seven grandchildren. My wife, Elizabeth, was born in the Philippines of an American mother and Dutch father, and so introduced into our marriage both European and Asian roots. We met in Cambridge, Massachusetts, where she received her doctorate in Asian Art History at Harvard University. In addition to raising our children, she taught art history for many years at Boston University, rising to a full professorship there. Elizabeth brought into our marriage a very welcome appreciation of the arts. She is also a good friend to my two children whom she met when they were teenagers.

Elizabeth and I were married in 1977. In the first several years of marriage, and then much later, after the children were on their own, Elizabeth was able to travel with me on my overseas professional activities, notably to Iran in the late 1970s, and to Singapore for an extended period in the late 1990s. In the intervening years, we spent much of our free family time at our second home, a white house in the Berkshires of Massachusetts that was our version of the old "family farm" of generations back. We also as a family found many occasions for overseas travel, including the sabbatical year in Japan in 1989 and another sabbatical in France in 1996. Most of the time, however, Elizabeth and I have lived where, we each knew from our college student years, we wanted to spend our lives: Cambridge, Massachusetts.

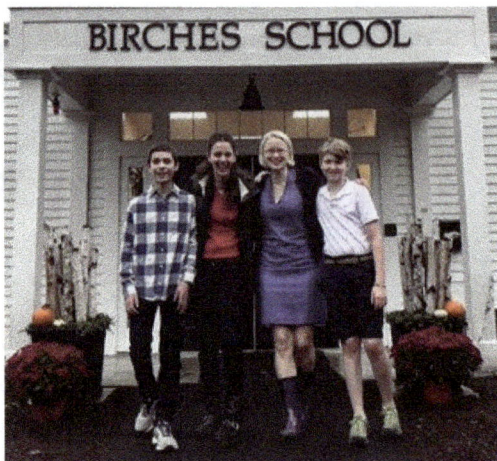

Fig. 3.24 This photograph from 2018 shows the Birches School co-founders. Cecily and her son Dexter stand at the right.

In our later years, Elizabeth and I have found much joy in contributing outside our professional areas in the formation and conduct of two non-profit organizations. We were founding members of the board of Yo-Yo Ma's "Silk Road Project" for the 20 years in which he led the organization, from 1998 to 2018, serving n various capacities during that time. It was wonderful to contribute in our small ways to the cross-cultural explorations of global music, resulting in the uplifting of the hearts of so many throughout the world. In 2012, our daughter Cecily and a friend started Birches School, a nature-based Kindergarten through 8th Grade school, in the basement of a church in Lincoln, MA, with an initial enrollment of five students. Elizabeth took on the "temporary" position as Head of School, a role she would fill for the next six years, and she and I both assumed board roles. Today the school enrollment is nearing 100 students. It now exists on the former property of a computer entrepreneur Dr. An Wang in a building that had been his home, with extensive renovation and expansion. (Fig. 3.24)

3.2.7 *Concluding remarks*

I write this at the age of 91, grateful for the years I have had and for the opportunities that came to me along the way. I have been fortunate to have

Fig. 3.25 A 90th-birthday celebration given me by a group of former graduate students. Elizabeth and I are at the center. Co-hosts David and Ginni Spencer are to our right and co-hosts Diran and Seta Apelian are to our left.

found a profession that I loved and so to have developed the passion necessary to succeed. I remember, from high school days on, being ambitious to make a contribution in my lifetime and to leave behind a body of work of lasting importance. With that as the primary aim, I presumed financial security would follow.

I was to learn the joys of discovery not only in the laboratory, but also in being with friends, students and family outside the laboratory. I have learned and experienced much more of the world than I could have dreamed of in my youth. I have certainly learned that lifetime friends are the best kind. I was deeply moved when a group of my former graduate students, mostly from the 1970s, gave me a surprise 90th birthday party (Fig. 3.25).

The world is a good deal more complicated than it was 70 years ago when I finished my undergraduate degree. Then, the Second World War was over and the country was hard at work building a prosperous peacetime economy. There were political issues and some fear of nuclear war, but, overall, optimism for the future was everywhere. For the graduating classes today, the picture is different. Industrial and economic successes in my lifetime have exacerbated old problems and brought new ones. Problems we did not imagine 70 years ago include global pollution of all kinds and the now recognized existential problem of global warming. Still, I think a good recipe for life today is as it was 70 years ago: to find a way to make a contribution, with a loving partner to share the journey.

Chapter 4

Yiu-Wing Mai – from Hong Kong to the world through composites science

4.1 Introduction by the Editor

Composites are a class of materials. A composite is a solid consisting of a number of constituents, such that the combination of the constituents enables performance that cannot be attained by a single constituent. An example of a composite is one with continuous carbon fiber held together by epoxy (a polymer). The fiber serves to reinforce the composite, while the epoxy serves as a binder (known as the matrix). The fiber is stiff and strong, but it is brittle. The matrix is weak but relatively ductile. As a result, a crack trying to propagate through the composite material tends to be blunted when it meets the matrix, so that the composite is less brittle than the fiber itself and is stronger and stiffer than the matrix itself. Because of the low density of both carbon fiber and epoxy, the composite is lightweight and is thus used for aircraft and sporting goods (such as tennis rackets).

The design of a composite material involves numerous considerations, specifically the choice of the constituent materials (e.g., the particular fiber and the particular matrix), the shape and dimensions of each constituent material (e.g., the shape and dimensions of the fibrous or particulate reinforcement, with the dimensions ranging from the nanoscale to the macroscale), the number of fibers in a fiber bundle (known as a tow, with thousands of fibers in a tow), the proportion (typically expressed as the fraction by volume, i.e., the volume fraction) of each constituent, the spatial distribution of each constituent (e.g., the degree of uniformity of the spatial distribution the reinforcement in the composite), the degree of continuity of the fiber (with short fibers being typically not as effective as continuous fibers for reinforcing the composite), the orientation of the fiber (with the composite strength being higher in the fiber direction than the direction transverse to the fiber direction), the use of multiple fiber orientations (in order to tailor the strength of the composite in various directions), the weave pattern (in case that the fiber is in the form of a woven fabric), the degree of fiber alignment for each fiber orientation (as the alignment is not perfect), the degree of bonding between the constituent materials (such as the bonding between the fiber and matrix), the degree of infiltration of the matrix to the space between the fibers in a tow during composite fabrication (with the incomplete infiltration resulting in pores, which are detrimental to the mechanical behavior), the degree of bonding between the fiber plies in a laminate (with each ply, known as a lamina, consisting of a large number of fiber, the laminate comprising a stack of the plies, and the interface between the adjacent laminae being typically the weak link in the composite), and the porosity in the composite (i.e., the volume fraction of the pores).

The design of a composite requires the calculation of the properties (such as the strength) of the composite based on the information (e.g., strength, volume fraction, etc.) for each constituent. Such calculation requires knowledge of the scientific relationship between the structure and properties of the composite. In case of the properties being the mechanical properties, the science is often referred to as engineering mechanics. The mechanical properties include the strength, stiffness (known as the modulus, which is the stress divided by the strain, where the stress is the force divided by the cross-sectional area in case of tensile or compressive loading, and the strain is the fractional change in the dimension), ductility (the strain at failure), and toughness (energy needed to fracture a unit volume of the material, as given by the area under the stress-strain curve up to fracture). The fracture energy is akin to the toughness, but it is different, as it is defined as the energy required to open a unit area of the crack surface.

The design needs to be checked by experimental work that involves the fabrication and testing of the composites. In addition, experimental work provides information (such as the mechanical properties of each of the constituent materials and of the interface between the constituents) that helps the design.

Professor Mai is a leading international expert in the use of engineering mechanics for the design and understanding of composite materials. He has received numerous significant honors, particularly in Hong Kong (where he grew up), China, the United Kingdom and Australia (where he is a professor in the University of Sydney). From a humble beginning, his life experience is rich and inspirational. Below is his life story in his own words.

4.2 Life experience as told by Professor Mai

In memory of my late mother, Ms. Yuet-Yau Tsui (1920-2021)

4.2.1 *My family background – From Qingyuan (清遠) to Hong Kong*

When I was small and learning to write my name in Chinese characters, I was so happy with my family name (米) which is so easy (for it is just like the national flag of the United Kingdom), but I felt awesome with my given names (耀榮) which are so complex comprising nearly all the eight basic brush strokes in Chinese calligraphy. It did not take long for me to realize that my family name (米), meaning rice, and pronounced as Mai (in Cantonese) or Mi (in Mandarin), is quite rare. In fact, through primary school to high school and to university in Hong Kong, I never met another classmate with the same family name 米. This has sparked my strong desire and curiosity to find out more about my ancestors and my roots, which have been facilitated much later in my frequent visits to China since 1983. The fact that there are only ~0.5 million people surnamed 米 in China clearly helps.

Qingyuan (清遠) is a prefecture-level city in northern Guangdong Province with a beautiful Bei Jiang (North River), known for its Small Three Gorges (小三峽) scenery, flowing through and dividing it into two parts (Fig. 4.1). It is the hometown of Zhu Rujen (朱汝珍) who was the Bangyan (榜眼) in the last imperial court exam of the Qing Dynasty in 1904. My father, Mai (also Mi) Lam, and mother, Tsui Yuet-Yau, came from adjacent villages, Hefen (河汾) and Shangcun (上村), about 30 km southwest from

Fig. 4.1 Beijiang (North River) flowing through Qingyuan City.
碧水蓝天北江清远美如画
(李作描 摄) (http://www.gdqy.gov.cn/jjqy/ljqy/content/post_1472259.html).

the city centre of the Qinxin (清新) District. My mother did not learn to read or write, but she was smart and clever, and could have received a good modern education like her sisters had she not been too shy to wear skirts to school. My father only received a couple of years of old-style education of basic Chinese classics in the village. My grandparents and many generations before them were farmers, owned some land for growing rice, and inherited a less than moderate house to live in (Fig. 4.2). To search for a better life, like many young people from the villages then, my father went to Hong Kong and began an apprenticeship in carpentry, especially piano-making, with the masters in Tsang Fook Piano Co. Ltd. owned by Mr. Tsang Fook who also came from Qingyuan. This explained the very nice furniture he could make for our homes in Hong Kong. After marrying my mother, he returned to Hong Kong and worked as a carpenter in shipbuilding for his brother-in-law who was at that time one of the few contractors in HK & Whampoa Dockyard or Kowloon Docks (Fig. 4.3) for short and commonly called the Big Factory (大廠) by the locals. My mother reunited with him after WWII and before the New China closed the border to Hong Kong in late 1949.

Life was truly difficult in the early 1950s in Hong Kong. Jobs were scarce and many refugees arrived from various parts of China rapidly increasing the population to over 2 million. Housing was a main problem.

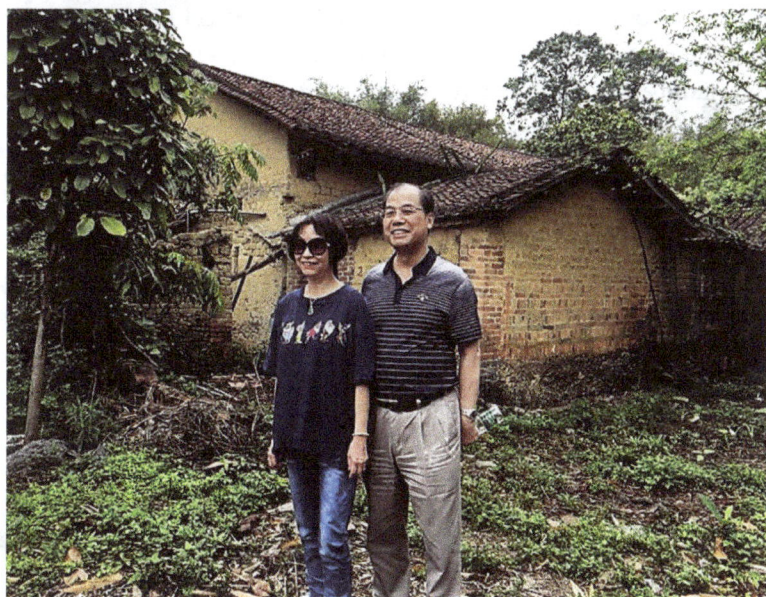

Fig. 4.2   Ancestral home in Hefen village, Qingyuan. Somehow, we still feel very much connected to it.

Fig. 4.3   Hong Kong & Whampoa Dockyard (also known as Kowloon Docks) in 1950s.
(https://industrialhistoryhk.org/indhk/)

I remember we lived on the top floor of a 4-storey red-brick Tong Lau (唐樓) at the corner of Chatham Road and Wuhu Street in Hung Hom in 2 small rooms divided by thin wood veneers. On the same floor, there were 6 other families living in 2 partitioned rooms and several bunk beds in the balcony shielded by hung cloths at night. All the tenants came from the same neighboring villages in Qingyuan and knew each other. The bathroom, "toilet" (with no water flushing then), and kitchen (with individually owned firewood stoves) were, however, for communal use. Living conditions were crowded and sometimes unhygienic, but we lived quite happily there despite occasional frictions and quarrels between tenants. When the family size grew to 3 boys and 2 girls (Fig. 4.4), we rented a flat nearby in Station Lane not far away from the Hung Hom Kwun Yum Temple (紅磡觀音廟). Eventually, my father saved enough money to buy a small flat in a then new housing project (Whampoa Estate - 黃埔新邨) within the compound of Kowloon Docks in early 1970s by which time shipbuilding had fast disappeared in Hong Kong. I did not realize that my mother had lost the eldest daughter before me in an accident while working in the rice-field many years before. I only learned about this much later from my wife.

My father did not know and could not tell me much about our ancestors or genealogy. Anyway, for his minimal education, he would not have the knowledge to do so. He only mentioned that we are the direct descendants of Mi Fu (米芾) and his son Mi Youren (米友仁) as proudly displayed in

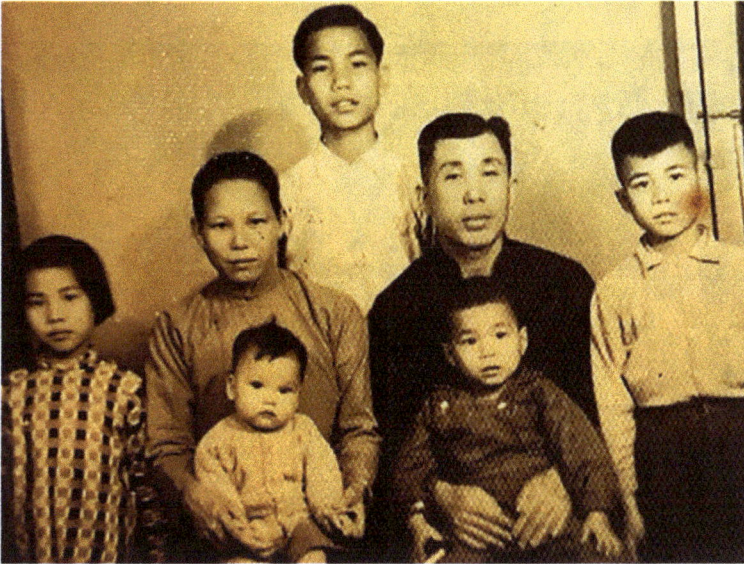

Fig. 4.4 Mai family photos in 1959. I am standing at the back. (L to R): Sisters Chung-Sum, Chung-Ho on mother's lap, brothers Yiu-Shing on father's lap and Yiu-Shang.

the entrance of the Mi ancestral hall in Qingyuan. (Fig. 4.5). Both father and son, particularly Mi Fu (Fig. 4.6), were renowned for their unique style of calligraphy and their innovative creation of the famous "Mi dots" (米點) technique for cloudy mountains in Chinese landscape painting in the Song Dynasty. In painting, the "Mi dots" technique predated the similar pointillism technique (using colour dots) developed by the French impressionist painters Georges Seurat and Paul Signac by more than 800 years. Stippling, invented much earlier than pointillism by the Italian painter Guilo Campagnolais in 1510, is a similar technique but uses solely black dots. Mi Fu grew up in Xiangyang (襄陽) in Hubei Province. Three brothers of the 10th generation descendants moved from Xiangyang to Qingyuan to start a new life and a branch of the Mi family over 700 years ago during the Yuan Dynasty. I am the 27th generation descendant. Maybe it is in the family DNA, I am quite good in Chinese calligraphy (Fig. 4.7). Many may not know but there are actually two Mi Fu Memorial Halls (米公祠): one in Xiangyang (Fig. 4.8a) and one in Wuwei (無為, Fig. 4.8b), Anhui Province. In addition, there are also a Mi Fu Calligraphy National Park (Fig. 4.9 L),

Fig. 4.5   Mi (or Mai) family ancestral hall in Qingyuan.

Fig. 4.6   Images of Mi Fu: (L) As depicted by Chao Buzhi (晁補之) in 1107; Public
Domain https://en.wikipedia.org/wiki/Mi_Fu#/media/File: 米 芾 .jpg; and (R)
Self-portraithttps://tangshi.tuxfamily.org/mifu/

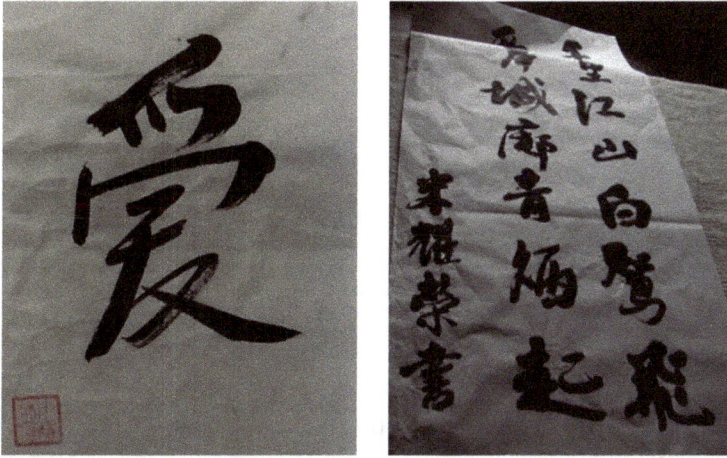

Fig. 4.7   My Chinese calligraphy.

Fig. 4.8   Ancestral Halls in (a) Xiangyang, Hubei Province and (b) Wuwei, Anhui Province.

Fig. 4.9   L: Mi Fu Calligraphy National Park, Zhenjiang. R: Mi Fu Square, Zhenjiang, Jiangsu Province.

which is unique in China, and a Mi Fu Square (Fig. 4.9 R) in Zhenjiang (鎮江), Jiangsu Province. I am proud to have visited all four landmarks.

4.2.2 *Growing up in Hong Kong and school days*

It was a unique experience growing up in 1950s in Hong Kong. There were no luxuries of toys (unaffordable) and electronic games (not yet created). My father worked in the dockyard and my mother helped to make ends meet by taking piecework home, such as removal of overflow flashes from injection molded plastic flowers, besides caring for my younger siblings. They did not have time for me. Therefore, I had my pastimes roaming the streets and learning their dark secrets, watching street shows, catching tiny crabs on the beach, playing marbles and soccer on sandy grounds, etc. Among these games, I particularly enjoyed flying kites. The flat rooftop on the Tong Lau we lived in was an ideal place to fly kites high in the open sky. There were many such rooftops in nearby Tong Lau. Kites were made of strong papers mounted on a frame made of bamboo splints and secured to cotton or nylon lines in a spool with two long handles. Kite fighting or duelling is to reign in the sky by cutting other kites off their lines! For this purpose, sharpened abrasive lines are required in addition to the spade-shaped kites (with no tails) that are easy to maneuver at the spool. Such lines were made by passing them through two holes of controllable size on a match box containing wet rice (or glue) and ground glass from broken thermos flasks and allowed to dry under the sun. Wet rice (or glue) provided the glass powder strong adhesion onto the cotton/nylon lines and the cutting function. That was my first experience of fabrication of functional composites! The Great Wall of China was thought to be built using sticky rice to bond the bricks together.

4.2.2.1 *Primary School – Kowloon Docks Memorial School (KDMS)*

All the sons of relatives from my father's village and those nearby followed their fathers to work in the three trades (三行) of carpentry, cement/mortar rendering, and painting in their early teens. I did not know if my father wanted me to follow him as a carpenter or if he had other plans for me. In fact, it was my first cousins who took some time to convince him that I should go to school and receive an education. So, I started quite late in school compared to many of my classmates. In 1952, I went to a small local school, Ming Tak Primary School (明德小學), in a tenement house across Chatham Road from where we lived and by the side of a textile dyeing and finishing factory (called Tai Hing Dyeing Company to my best recollection), to prepare for the entrance exam of a better-equipped government school, Kowloon Docks Memorial School (KDMS) (九龍船塢紀念學校) to which I was admitted in 1953. KDMS (Fig. 4.10) was established in 1949 in memory of those workers who died in the dockyard during WWII by Japanese bombs. Most students were from families of workers employed in Kowloon Docks. Girls in this school had to study a subject on homemaking including sewing and knitting. Boys took on woodwork and learned to use wooden planes, groove cutters, chisels and drills, shellac finish, lacquer

Fig. 4.10 My Primary School - Kowloon Docks Memorial School (KDMS)
(https://lausoldier.blogspot.com/2014/11/blog-post_65.html)
(https://www.flickr.com/photos/50066665@N03/4790476714)

coatings, and cow skin glues. These would be my first exposure to tools for carpentry although I had seen them all and how they functioned from my father who made the furniture for our homes. I fully enjoyed my 6 years in the school and graduated in 1959 (Fig. 4.11). I have formed firm friendships with some classmates whom I still meet, after more than 6 decades, during my visits to Hong Kong, reminiscing the good old days when we were once young, carefree and innocent (Fig. 4.12). Sadly, the school was

Fig. 4.11   KDMS 10th cohort of Primary 6 graduates and the teachers on June 24, 1959. I am standing in the back row, 12th from the left.

Fig. 4.12   Kowloon Docks Memorial School (KDMS) classmates. I am 3rd from the right standing.

closed in 1995 by the Education Department due to insufficient student enrolment. It is a shame that it has been unused since then and left in ruins today.

In the 1950s and 1960s, owing to the shortage of classrooms, schools were often used to run a morning session and an afternoon session to accommodate more students. In this context, it is interesting to note that KDMS was an AM-session school, and the campus was used by the Hung Hom Kaifong Association Primary School (紅磡街坊公立學校) as a PM-session school. In the latter school, then a young teacher, Szeto Wah (司徒華), taught there from 1952 to 1961 immediately after graduation from the Grantham

Fig. 4.13 My secondary school - Victoria Technical School (VTS) – Red Brick House.

College of Education. Later, he rose to become the most noted democracy activist and influential politician in Hong Kong.

4.2.2.2 *Secondary School – Victoria Technical School (VTS)*

I sat the Joint Primary 6 Examination in 1959 which was administered by the Education Department responsible for assigning students to different secondary schools in Hong Kong. I was supposed to go to Clementi Middle School but my father was very keen for me to attend a secondary technical school believing that I would not starve with some trade skills. This was easy since many would prefer a secondary grammar school and hence I was quickly re-assigned to Victoria Technical School (VTS) (維多利亞工業學校) which was commonly known as the Red Brick House (Fig. 4.13) in Wood Road, Wan Chai. VTS did not offer biology, botany, history or English literature but did cover most subjects as other grammar schools. The distinct feature would be the compulsory technical subjects on geometrical and machine drawing, woodwork (or metalwork) theory & practice, and technical knowledge. Little did I know then that these technical subjects, especially geometrical and machine drawing, would give me a head start when I did mechanical engineering at the University of Hong Kong some years later. Indeed, this second experience working with wood enabled me to understand more about their directional-dependent properties, longitudinal splitting *versus* transverse cracking, and environmental effects on wet (tough) wood *against* dry (brittle) wood. Naturally, in those days, I would not have understood the terms of *anisotropy* (that is, different physical and mechanical properties along different testing axes) and *fracture energy* (that is, the work required to create a unit area of new surfaces) as two important attributes of wood as a unique composite material. But wood comes from trees. As Prof. Charles Gurney (later my PhD supervisor) lamented that new strong materials are made by fools like him but only God can make a tree!

On the days when we had woodwork or metalwork practical classes, we would wear blue overalls to school. I vividly remember walking from home in Hung Hom to take the Jordan-to-Wan Chai ferry, and from the pier, crossing several main roads like Jaffe Road. Lockhart Road and Hennessy Road, passing a number of night clubs and dancing halls on the way, and through Heard Street arriving at the school in Wood Road in blue overall attracted quite a few staring eyes. VTS was a very good school in my time (1959-1966). There were 5 classes from Form 1 to Form 3 and then 3 classes from Form 4 to Form 5. Hence, many Form 3 students would not be promoted to Form 4 and became apprentices in different industries or continued in other high schools. Some of them, in fact, had very successful careers and went on to become leaders in their professional fields, supporting the manufacturing industries and contributing to the fast economic growth of Hong Kong in the 1970s through to the 1990s. Back then, I liked and did better in arts than science subjects. I was accepted by Kowloon Wah Yan College to study Lower 6 arts after the English School Certificate Examination in 1964. My father was visibly not too keen and I finally gave in to continue at VTS to do Lower 6 and then Upper 6 science. Nonetheless, I did well enough in the Advanced Level math (pure & applied) and science (physics & chemistry) subjects, matriculated and was accepted in 1966 to do engineering at Hong Kong University (HKU). This was the first cohort of 23 Upper 6 students presented by VTS for the HKU Advanced Level exams. We had 19 students matriculated, 8 went to engineering (which were more than 6 from King's College (英皇書院) in a first-year intake of 60), 4 to science, 1 to architecture and 1 to social science at HKU as well as another 3 went to US and Canadian universities. That was an exceptional cohort and these brilliant results were not to be repeated. I still see some of the high school classmates whenever I visit Hong Kong (Fig. 4.14). Victoria Technical School changed its name to Tang Shiu Kin Victoria Government

Fig. 4.14   Reunion dinner with VTS classmates. Seated: 1st left (Marco M-H Wu, former Director of Buildings HKG), 2nd left (me). Standing: 2nd left (Roger S-H Lai, former Director of Electrical & Mechanical Services HKG) and 3rd from right (Kenneth T-W Pang, former Commissioner of Rating & Valuation, HKSAR).

Fig. 4.15   University of Hong Kong.

Secondary School in 1997 when it moved from Wood Road to Oi Kwan Road. It is now co-educational and is no longer the same school I went to.

4.2.3 *First generation of family to attend university*

The University of Hong Kong (Fig. 4.15) admitted less than 10% of matriculated students into its various degree programs. It was therefore extremely

Fig. 4.16   (L to R) Peel Laboratory, Duncan Sloss Building and Ho Tung Engineering Workshop of the Faculty of Engineering in 1960s.

competitive to get in. However, most would have good rewarding jobs for life, especially in government departments, on getting out. My parents were truly proud that I would be the first-generation of the family to attend university. But they were clearly worried about the expensive costs involved since my younger brothers and sisters were still in school. Anyway, I managed the financial issues with the Hong Kong Government bursary and Wah Kiu Yat Po (華僑日報) scholarship, which were supplemented by incomes from private coaching and teaching matriculation classes in evening schools. HKU was long considered overly elitist since it was, for many years, the only university in Hong Kong. It should be remembered that the Chinese University of Hong Kong was not established until 1963 by combining the 3 existing colleges of New Asia College, Chung Ki College and United College. Also, no engineering degree programs were offered at that time. Many HKU students did come from the upper-class families; but in my time, it was slowly changing. The university admitted more qualified young men and women from less privileged background. Despite what we might criticize the education policies of the British Hong Kong colonial government, many like me were fortunate and thankful for the opportunities to access a university education, climb the social class ladder and achieve economic success.

Engineering was offered by the university since 1911 and, as said earlier, only 60 students were accepted in 1996 (Fig. 4.16). Let me digress a little here. Sir Sze-Yuan Chung (鍾士元) was a very influential politician in Hong Kong. He graduated first class honors in mechanical engineering in 1941. Then, he did a PhD in 1951 at Sheffield University with Prof. Herbert Swift on wire drawing and his published papers are occasionally still cited today. Sir SY played a crucial role to establish all three technical universities in Hong Kong. These are the Hong Kong Polytechnic University, the City University of Hong Kong and the Hong Kong University of Science and Technology. As with the British collegiate university system, all students at HKU must belong to a residential hall or college in order to be presented for exam and graduation. I belonged to Ricci Hall but I was non-residential. There were two other halls, Hornell (men) and Duchess of Kent (women), exclusively for non-residents. Exam results were posted on the Engineering Faculty noticeboard for everybody to see. Year 1 was common, but in Years 2 and 3, students had to choose a discipline from civil, electrical and mechanical engineering. In my case, I always feared electricity and was undecided between mechanical and civil engineering. I could not remember exactly how or why I ended up with mechanical engineering. It could be that I had the advantage of already learnt technical drawing at VTS. Or it could be the new arrival of a distinguished scholar, Prof. Charles Gurney, from Cambridge/Cardiff universities to HKU mechanical engineering that influenced me towards this discipline.

The subjects I enjoyed most and received rigorous training for three years were solid mechanics as they were taught by excellent teachers. I also liked thermofluids and did a final year thesis project with Mr. Samuel Wong (Fig. 4.17b) on the heat and mass transfer characteristics of a rotating disk filled with naphthalene at different speeds. However, materials technology and metallurgy were not taught and I only had limited knowledge on concrete. Composites were not even heard of. Three years went by quickly. I came second in Year 1, first in both Year 2 and Year 3, and received a first class honors BSc(Eng) degree. In addition, I was also awarded the most prestigious Williamson Prize as the graduating candidate who had the best academic record as an undergraduate across the three disciplines of civil, electrical and mechanical engineering. There were only 10 graduates in mechanical engineering (Fig. 4.17a). That was 1969 and I have since lost contact with many of them.

(a)

(b)

Fig. 4.17   (a) 1969 graduates in mechanical engineering at HKU. I am 1st from the right in the back row. (b) Graduation dinner with the academic staff. I am standing 3rd from right; Prof. Charles Gurney seated in the middle front row. Also seated are Dr. Nelson Chen 1st left, Dr. Chi-Loong Chow 1st right and Mr. Samuel Ping-Wai Wong 4th from right.

I had thought of going to the US or the UK for postgraduate studies but it would be good to work for a few years first and saved some money. However, a job interview with Mark Wong, the Manpower Division Head of the Hong Kong Productivity Centre (HKPC), changed all this plan. He

Fig. 4.18   My PhD supervisor, Prof. Charles Gurney.

created a generous 3-year HKPC scholarship for me to do a PhD in HKU on condition that I would join the Centre as a Management Technology Trainer after completion. This was an attractive offer. However, there was not really much research in mechanical engineering at HKU then. I was also not prepared. For I did not even know what research topic to undertake and who should be my supervisor. In the end, Prof. Charles Gurney (Fig. 4.18) came into the picture. He suggested that I worked on the fracture mechanics of materials and also served as a department demonstrator (similar to a TA in the U.S.). I did not know that he was involved in identifying that the failure of the De Havilland Comet in the early 1950s was due to metal fatigue of the square windows. Having worked with Dr Alan Arnold Griffith, generally recognized as the father of fracture mechanics, in the Royal Aircraft Establishment at Farnborough in the 1930s to 1940s, Gurney was an enthusiastic supporter of the energy approach to fracture mechanics as elegantly elucidated by his seminal paper (*Proc. Roy. Soc.* A299, 1967, 508). He gave me this paper to read on our first meeting. It not only formed the physical basis of my doctoral research but also ignited my continuing interest and passion on the subject. Interestingly, many years later, I did my postdoc work with Prof. Tony Atkins (University of Michigan) and Prof. Gordon Williams (Imperial College London) (Fig. 4.19) who are both enthusiasts of the Griffith-Gurney energy approach to analyze the fracture of materials in the elastic, elastoplastic and plastic regimes.

In those early years in HKU mechanical engineering, the research culture was weak, facilities were somewhat primitive and lacking, relevant journals were unavailable in the library, and there were very few graduate

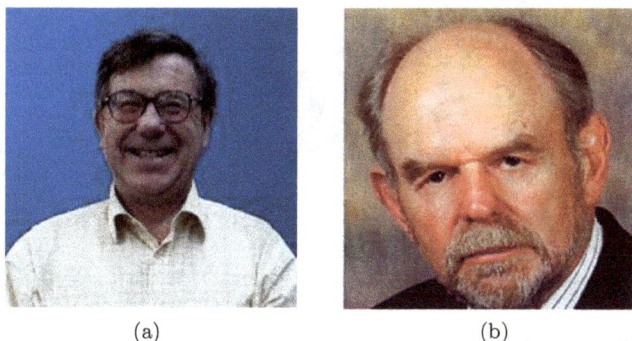

(a)   (b)

Fig. 4.19   My postdoc supervisors. (a) Prof. Tony Atkins *FREng* (University of Michigan). (b) Prof. Gordon Williams *FRS FREng* (Imperial College London).

students. My doctoral thesis project was on the rate of slow cracking of materials and, despite the adverse learned environment, I did some good original work. One small project I worked on the side with Robin Owen, an assistant lecturer in metallurgy and MPhil student, was of particular interest. In geometrically similar structures of materials with high *fracture toughness* (defined as the material resistance to rapid propagation of a crack under an applied stress) and low *yield strength*, which is akin to having large crack-tip plastic zones, cracks occur before yielding in large structures but *vice versa* in small structures. This means that small size specimens tested in the lab give ductile failure and cannot reproduce the brittle failure often found in large size specimens – a manifestation of the size effect on fracture. This observation agrees with our experience in comminution of solid particles by compression (or crushing) in that there is a brittle-to-ductile transition at a critical size below which the particle cannot further comminute. We overcame this difficulty using a specially designed test rig consisting of 4 steel U-sections containing an I-section specimen (Fig. 4.20) whereby the plastic zone was confined at the crack-tip region as the crack propagated. The U-sections remained elastic throughout the process. These results are reported in my first Royal Society paper (*Proc. Roy. Soc.* A340, 1974, 213). When I finished my thesis in 1972, I was only the 3rd PhD student to graduate in mechanical engineering. Gurney was a great mind, a poet, an engineering physicist. His main achievement in the field of fracture mechanics was the development of a full body of knowledge on crack propagation based on the energy approach and the first and second laws of thermodynamics to

Fig. 4.20 Fracture of high toughness/yield strength ratio materials. (L) Test rig showing I-section test sample slotted vertically at the center of the 4 bolted high strength steel U-sections under applied bending moments. (R) In the laboratory with Robin Owen (ca. 1971).

understand crack resistance, crack stability, crack path, etc. I was his last PhD student in Hong Kong and he taught me ways to original thinking, self-learning, etc. among other things. He and his second wife, Sophie Gurney (a member of the 21 Group of Artists and a great-granddaughter of the naturalist Charles Darwin) moved to an old village, Harbertonford, near Totnes, Devon after retirement from HKU in 1973. My wife, Louisa, and I visited them in early 1980s when I was on sabbatical at Imperial College London working on the fracture mechanics of shallow cracks in uPVC water pipes. Gurney showed me the design of a bridge he wanted to build across the river at the back of his cottage. Louisa taught Sophie Gurney how to play mah-jong. Sadly they are no longer with us.

Now was the time I had to fulfil my obligations to the HKPC which supported me with a generous PhD scholarship. However, I quickly realized that it was not the place for me to develop a career and I resigned after 6 months. This decision changed my life completely.

4.2.4 *Journeys to the U.S. and the UK – Research on composites, ceramics and polymers*

In 1973, Prof. Bernard Crossland of Queen's University of Belfast (QUB), famous for explosive forming and welding work, visited Gurney in HKU as an external examiner for our mechanical engineering degree program.

He offered me an Assistant Lecturer position. This was at the height of the sectarian war in Belfast. Whilst waiting for the work permit to arrive, I decided to go to the University of Michigan in Ann Arbor as a postdoc for Prof. Tony Atkins who was Gurney's undergraduate student at Cardiff and did a PhD thesis on hot hardness of metals with Prof. David Tabor at the Cavendish Laboratory at Cambridge. The near-2 years with Atkins were most fruitful and they opened up my knowledge horizons.

I worked on two projects. One project supported by NASA was on the control of composite interface bonding to achieve both high fracture toughness and high tensile strength which are usually mutually exclusive. Fiber composites whose fracture surfaces exhibited long fiber pull-out lengths absorb a lot of fracture energies (Fig. 4.21a) and are tough. Conversely, a near-flat fracture surface with very short fiber pull-out lengths is brittle. Strong interfaces give short pull-out lengths and more effective stress transfer from matrix to fiber. Weak interfaces give long pull-out lengths and less effective stress transfer to the fiber. A solution to this paradox is to apply an appropriate intermittent or full coating with designed properties along the fiber length (Fig. 4.21b) and to allow the full strength of the fiber to be reached through stress transfer over a sufficiently longer length embedded within the epoxy matrix (Fig. 4.21c). Hence, the effects with and without a full PVAL coating, corresponding to longer and shorter pull-out fiber lengths, respectively, on the impact fracture energy of carbon fiber-epoxy composites are distinctly displayed (Fig. 4.21d). This coating concept could be applied from the 1D fiber length case to the 2D inter-ply interface area situation with a perforated or full plastic thin film for interlaminar toughening against delamination on, for example, composite parts and components. The analogy is laminated glass displays and/or polymer-laminated steels although they are for different purposes. In fact, these two ideas of 1D and 2D coatings/thin films resemble today's work on hierarchical fibers and interleafs for toughening and enabling the functional properties of composites to be realized.

Another project was funded by GM (U.S.) Warren Technical Center to understand the thermal shock behavior and ranking parameters for crack initiation and crack propagation of a range of ceramic cutting tools based on carbides, nitrides and oxides. This work had a long-term influence on my later research about grain-bridging of coarse-grained ceramics and

(a)   (b)

(c)

(d)

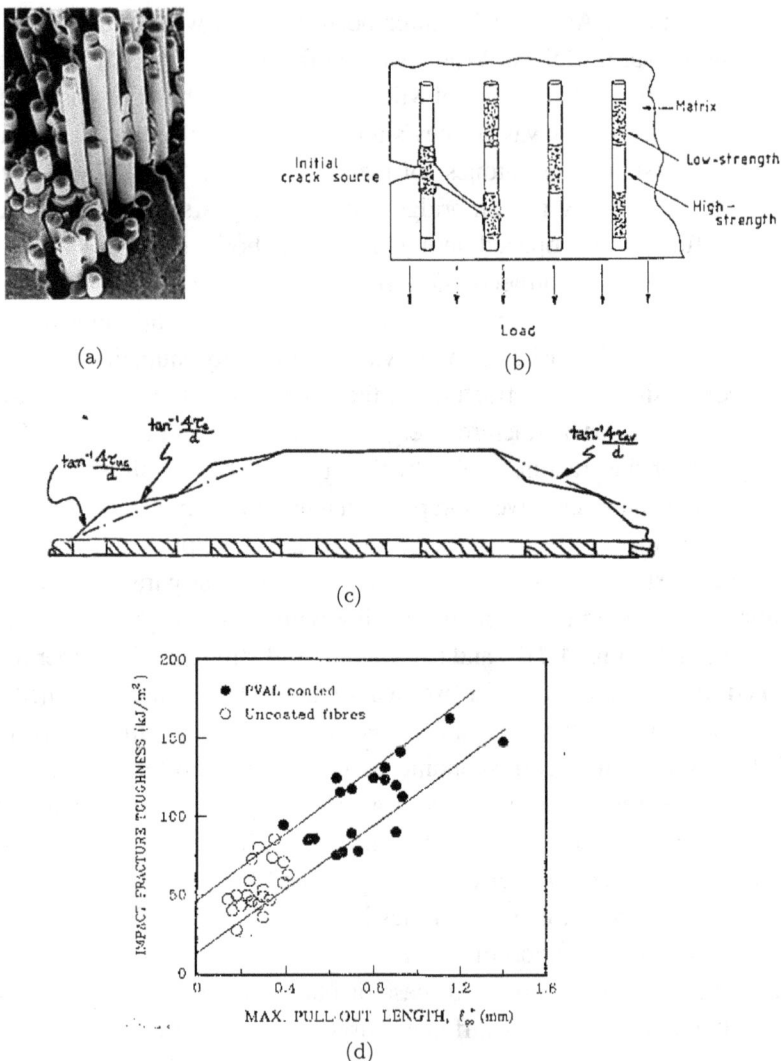

Fig. 4.21 (a) Fracture of CVI-SiC/SiC ceramic matric composites showing long fiber pull-out length. (https://en.wikipedia.org/wiki/Fiber_pull-out#/media/File:CVIpullout.jpg) (b) Intermittent coating along the fiber length with repeated coated and uncoated regions. (c) Stress transfer along intermittent bonded fiber length – more effective in uncoated and less effective in coated regions – ultimately reaching the full fiber strength. (d) Impact toughness improvement of CFRP with full PVAL coating on carbon fiber (CF) due to increased fiber pull-out length.

development of superhard nanocomposites coatings on steel tools for green machining.

Tony Atkins made an early reputation on the hot hardness measurements by the mutual indentation technique at the Cavendish Laboratory. He used the Griffith-Gurney approach for elastic-plastic fracture extending to the cutting of hard and soft solids. Atkins was a dynamic personality, energetic, imaginative and inspirational. He taught me the essential qualities of an innovative researcher: demonstrated enthusiasm with depth and breadth as well as proven capacity for both lateral and vertical thinking. He was also an authority on the Great Western Railway (U.K.). We wrote the first book on **Elastic and Plastic Fracture** (Ellis Horwood/John Wiley, 1985 and 1987) largely based on the Griffith-Gurney energy approach. I was shocked when I heard that he passed away on 25 September 2018 after a short but serious illness. Atkins was a great mentor and he gave me the first opportunity to do postdoc work at Ann Arbor opening the door of my academic pursuit.

I thought it would be great to have some British experience if I were eventually to return to an academic position in Hong Kong. So, I went to work for Prof. Gordon Williams as a postdoc in 1975 at Imperial College London on two projects. One was on the bimodal fracture mechanics analysis and molecular relaxation mechanisms of semi-ductile nylon and polyethylene. Another was on fatigue mechanics of polystyrene as affected by temperature and frequency using the Dugdale line-zone model and accumulated damage. These were a new and steep learning experience on a different class of materials which would be used as matrices for fiber reinforced composites. They would also open up new research directions for me in polymers and polymer blends (defined as comprising a mixture of 2 or more polymers which may be miscible or immiscible). Williams is known for his work on fracture mechanics of polymers and composites based on the energy approach. Many ISO and ASTM Standards on fracture in these materials are due to his research results. He taught me to first understand the physics of the problem, then formulate simple elegant mechanics models, solve for answers subject to given boundary conditions, and compare with experiments. He came to Sydney as a visiting professor for 3 months every year from early 2000s until he had a bad fall from a ladder in June 2018. He had not really fully recovered and has not been doing much scientific work since. A great loss to science!

(a)   (b)

(c)   (d)

Fig. 4.22   Research collaborators. (a) Brian Cotterell, (b) Brian Lawn *NAE FAA*, (c)
Shouwen Yu and (d) Klaus Friedrich.

The QUB work permit was finally issued but, by this time (mid-1976),
I already had firmed offers from Case Western Reserve University (through
Prof. Eric Baer), University of Hong Kong (through Prof. Chi-Loong Chow
who was my former teacher) and the University of Sydney (through a
phone call from Prof. Roger Tanner to Prof. Gordon Williams with whom
I was just discussing my research results in his office). As an aside, I found
out later that Tanner called because Dr. Paul Isherwood who was working
for Williams had accepted but then declined to go to Sydney. He stayed
at Imperial College London as a lecturer; and I was instead offered the
position he vacated. I eventually decided to go to Sydney. For I had, and
still have, great admiration and respect for Prof. Brian Cotterell (Fig. 4.22a).
He was an outstanding researcher from the Welding Research Institute and

Cambridge on fracture mechanics, especially crack path, and is extremely knowledgeable beyond his fields.

4.2.5 *Adventures to Australia – The University of Sydney*

When I arrived at the University of Sydney in early December 1976, I soon realized that the research facilities were scanty and somewhat dated. There were no graduate students in solid mechanics. Worse still, I did not see Brian Cotterell who was on sabbatical leave with Prof. Jim Rice at Brown University and he did not return until early 1978. It took us nearly a decade to establish a sizeable research group and set up good testing facilities funded by industry and government agencies. Cotterell is typically British, cool, and somewhat distant if you do not know him well. This probably explained why he never replied to me when I wrote to him for a postdoc position soon after my PhD in HKU and before I went to the University of Michigan at Ann Arbor. But he might also not have any funding. Anyway, it took me some time to win him over. In the end, we had a long productive research collaboration on a range of topics, notably essential work of fracture (EWF), crack tip constraints on ductile fracture and quasi-brittle fracture mechanics of cementitious materials and their fiber composites.

Essential work of fracture (EWF) for polymer thin films

Based on the energy separation approach, Cotterell already demonstrated a powerful test method to measure the specific essential work of fracture (same as fracture energy) of thin steel sheets in deeply double-edge-notched tension specimens. Realizing that this could help the polymer and paper packaging as well as newsprint (on runability) industries, we first extended the EWF concept to thin polymer films (*Int. J. Fract.* 32, 1986, 105). We verified that the simple methodology of partitions of the essential work (that is required to create new crack surfaces) and non-essential work (that is associated with plastic deformation unrelated to the fracture process) from the total fracture work could be used to obtain a true material fracture energy value which is otherwise difficult to obtain. This EWF idea triggered many research activities in different laboratories across many countries. For example, the Japanese Standards Association, the Pulp and Paper Research Institute of Canada, and the ESIS-TC4 Group in Europe. It has even been recently applied to measure the fracture energy, which helps the process

Fibre cement sheet manufacture using the Hatschek process

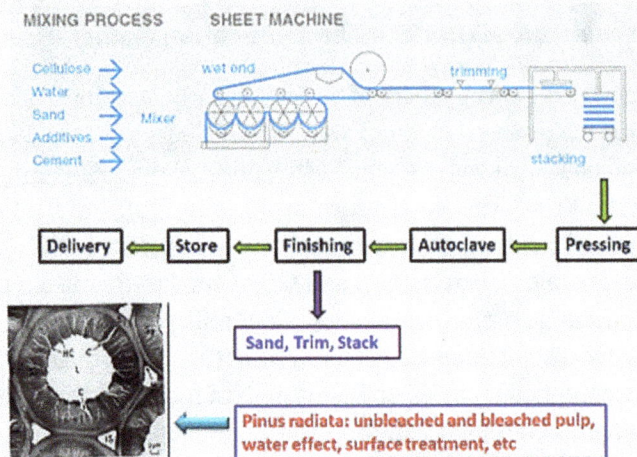

Fig. 4.23 Hatschek process for asbestos-free fiber cements. Cellulose fibers from *pinus radiata*. (https://www.jameshardie.com.au/fibre-cement)

design, of 3D printed short-fiber reinforced polymer films (*Compos. Sci. Technol.*, online 12/3/2022, 109361). Refinements of the EWF technique and analysis are also just reported (*Eng. Fract. Mech.* 260, 2022, 108442). An ISO/Draft International Standard (ISO/DIS 23524(en) *Plastics — Determination of fracture toughness of films and thin sheets: the essential work of fracture (EWF) method* – ISO/TC 61/SC 2, dated 2022-01) has been prepared for approval. This is solid proof of the global impact of the EWF concept on engineering.

Development of asbestos-free cements – James Hardie & Coy. Ltd.

Asbestos cements were widely used through 1950s to 1960s for buildings, heat shields, and drainage/water pipes among other applications. But asbestos fibers were known to cause lung cancer and mesothelioma. James Hardie Industries in Australia was the major manufacturer of asbestos-based cement products and was required to phase them out by late 1980s. The search was to source a replacement fiber for asbestos which would give comparable mechanical properties and also be compatible with the Hatschek process used (Fig. 4.23). Cellulose fibers from *pinus radiata* were

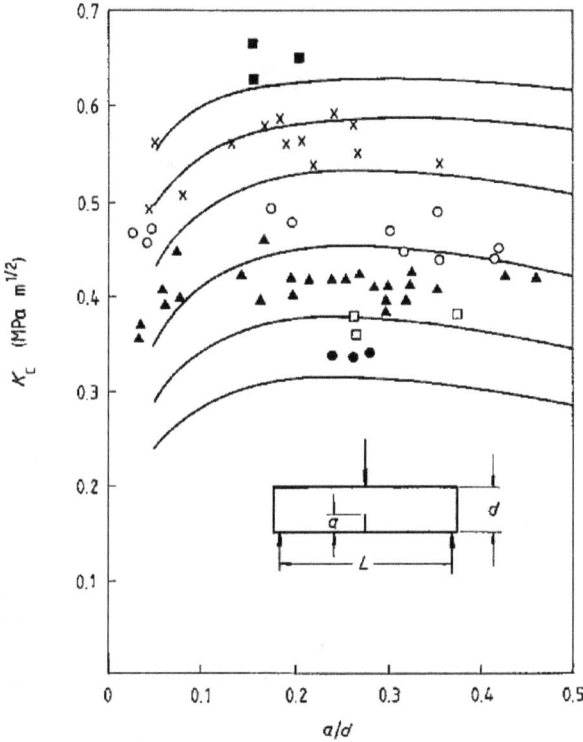

Fig. 4.24 Fracture toughness Kc as a function of initial notch length for cement pastes of various specimen size with L = 5d. d (mm) = (•) 5, (□) 8, (▲) 14, (○) 28, (x) 56, (■) 110.

finally selected and a wide range of new asbestos-free cements were successfully produced and used in Australia and many other countries. This was our first large industry-funded project with obvious benefits to society and we were happy to have contributed to this cause. We were in fact more pleased with our basic contributions to the knowledge base of a then new research area of *quasi-brittle fracture mechanics* which applied to cementitious materials and fiber cements. Even for cement paste, a one-parameter characterization using a *critical stress intensity factor* K_C (also called fracture toughness) based on the initial notch length (a/d) is inadequate because it shows a strong dependence on the geometric size of the beam (L) (where $L = 5d \cdot 5 \leq d(\text{mm}) \leq 110$) (Fig. 4.24). In concretes and fiber cements, the crack-wake bridging effect caused by the aggregates and the fibers must also be considered (Fig. 4.25a). For this gives rise to the crack

(a)

(b)

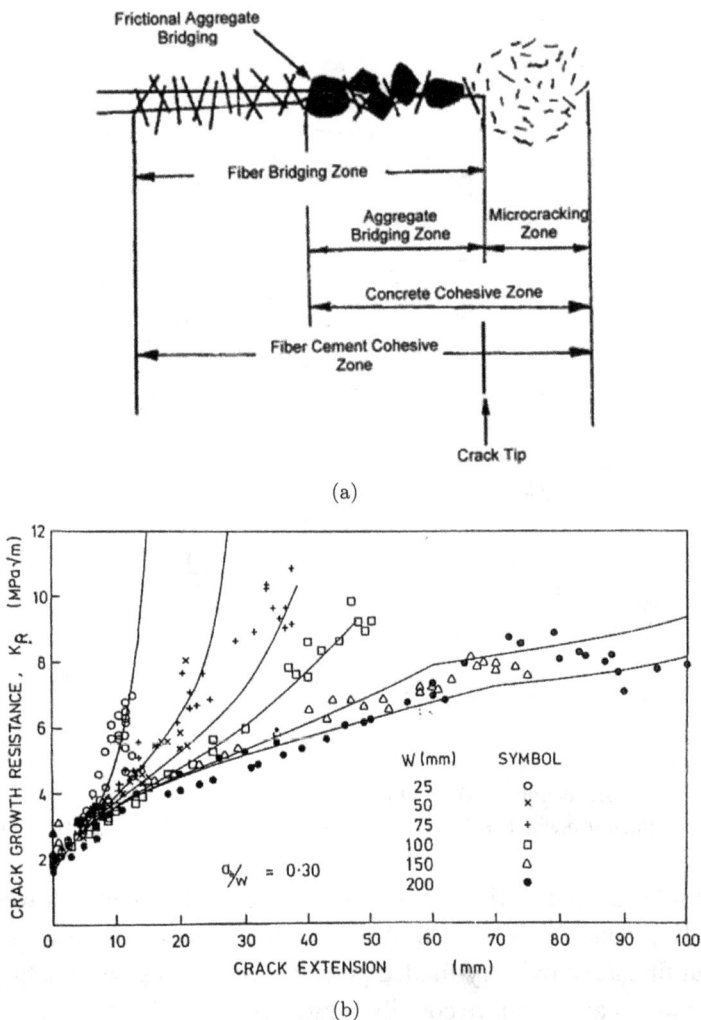

Fig. 4.25   (a) Bridging effects in cementitious materials and fiber cements. (b) Size effect on crack growth resistance (R) curves of fiber cements from geometrically similar notched bend specimens with different beam depth W and fixed a_0/W ratio of 0.3.

growth resistance (R) curve. That is, the fracture resistance (expressed in either fracture toughness or fracture energy) increases with crack growth. Unless very large specimens are tested, size-dependent R-curve behavior will be obtained (Fig. 4.25b). Cotterell and I wrote probably the first book on *Fracture Mechanics of Cementitious Materials* (Chapman & Hall, 1996).

Crack-wake bridging model of coarse-grained ceramics

I spent a sabbatical in 1985 and then 6 months in 1988 as a senior Fulbright scholar with Prof. Brian Lawn (Fig. 4.22b) in the National Institute of Standards and Technology (NIST) at Gaithersburg, Maryland to work on experiments and modelling of the effects of crack-wake grain-bridging on fracture and fatigue of coarse-grained alumina. The Mai-Lawn bridging model is in fact not very different to, and actually simpler than, that of the fiber bridging model for cementitious matrices. Lawn is a pioneer on indentation fracture mechanics applied to glasses and ceramics. He has great intellectual depth who taught me about critical thinking. The time in the NIST produced some most memorable work on crack-wake bridging in coarse-grained ceramics that still stands today (*J. Am. Ceram. Soc.* 70, 1987, 279 & 289). I remember the many after-lunch walks we had in the woods of the NIST compound and the jokes he cracked about a physicist, an engineer and a mathematician or a lawyer. In return, I recited to him many wise Chinese sayings and proverbs beyond the professions.

Evaluation of composites interface properties with fracture mechanics models

When a single fiber embedded with a given length in an elastic matrix is pulled at its end, it will start to debond around the circumference and progress along the fiber length. The fiber diameter is also reduced due to the Poisson contraction effect and there is relative frictional sliding at the debonded interface under a residual clamping pressure. We proposed a fracture mechanics approach by treating the debonding as a crack growth problem and obtained simple equations to evaluate the fiber-matrix interface fracture energy in the bonded region, the friction coefficient and the residual stress at the debonded region. This is the Gao-Mai-Cotterell model (*ZAMP*, 1988, 39 550). Various refinements and extensions of this composite interface model were made by several PhD students at the time including Jang-Kyo Kim, Limin Zhou and Hong-Yuan Liu. Later, Kim and I summarized all these results in a research monograph on ***Engineered Interfaces in Fibre Reinforced Composites*** (Oxford, Elsevier, 1998). I should acknowledge that Prof. Chun-Hway Hsueh (薛承輝) at National Taiwan University (then Oak Ridge National Laboratory) also worked on the interface mechanics of the single-fiber pull-out problem independently at about the same time. However, he used a shear-strength-based approach instead of our fracture mechanics approach. I visited him in Oak Ridge many years ago and he complained about having to drive 3-4 hours in the weekend to Atlanta for Chinese grocery! After he moved back to Taiwan, I visited him again in 2010 when I attended the ACCM-10 meeting in Taipei and I joked that he now did not need any long-distance driving for Chinese food. We have become good friends and I am also a distinguished research chair professor in his department.

In 1991, Brian Cotterell, finding Australian taxation burdensome, left Sydney to take up a professorship at the National University of Singapore. This created a big void in the materials and mechanics areas. However, it was soon filled by three new appointments (Fig. 4.26) to strengthen the

(a)

(b)

Fig. 4.26   (a) With Liangchi Zhang FTSE and Lin Ye FTSE; I am in the middle. (b) With Caroline Baillie.

expanding research in fiber composites including natural fibers (Dr. Lin Ye from IVW Kaiserslautern and Dr. Caroline Baillie from Surrey) and further develop the emerging field of nanotechnology (Dr. Liangchi Zhang from MITI National Mechanical Engineering Laboratory). I felt extremely proud that both Ye and Zhang were elected to the Australian Academy of Technological Sciences and Engineering (ATSE) on the first year of my nomination. It is interesting that, 3 decades later, they are both Chair Professors in Engineering at the Southern University of Science and Technology in Shenzhen. Also, Baillie has had an amazing career path and is now Professor of Engineering and Social Justice at the University of San Diego. Yes, it is *social justice* and is not a misprint!

4.2.5.1 *Turning point in my academic career*

I was offered in 1985 a tenured professorship by Prof. Chuck Vest in mechanical engineering at the University of Michigan where I first met him when I did my postdoc work and he was a young associate professor. Vest was then well-known for his work on holographic interferometry applied to a range of engineering problems. He was later to become the President of MIT. I was quite keen to go back to Ann Arbor for I had some fond memories of the place – not only its research excellence but also its sporting prowess. But my wife was hesitant due to the usually harsh winter in the mid-West. Therefore, we remained in Sydney, and in 1987, I was finally offered a second Personal Chair in engineering at Sydney. (The first personal chair went to Prof. John Booker in Civil Engineering in 1985). Vice-Chancellor John Ward also provided start-up funds to establish the Centre for Advanced Materials Technology (CAMT) for research focused on fracture mechanics and fiber composites. To strengthen these two fields, I had received immeasurable help from Prof. Shouwen Yu (Fig. 4.22c) of Tsinghua University on solid mechanics and Prof. Klaus Friedrich (Fig. 4.22d) of IVW, University of Kaiserslautern on fiber composites science and engineering. They recommended many excellent postdocs in the following years to the CAMT. Unfortunately, Friedrich passed away on May 29, 2021 and I have forever lost a true friend to whom I could always seek advice and guidance. Looking back, to stay in Sydney was the right decision. This was the turning point in my academic career.

Under the umbrella of the CAMT, we became a research partner of the Co-operative Research Centre (CRC) for Aerospace Structures (AS) in 1992 and Advanced Composite Structures in 1996 (ACS) covering the aerospace, land transport and marine industries. I was mainly working on improved manufacturing processes which included thermoplastic matrix composites, stitching and z-pinning, pultrusion and thermoforming of composite laminates. We made significant contributions in these areas, especially on the first mechanics models of stitching (with Dr. Lalit Jain who did a PhD with Prof. Robert Wetherhold at University at Buffalo, The State University of New York) and z-pinning (with Dr. Wenyi Yan, now full professor at Monash). In 1997, with new funding from the Defence Science and Technology Organization, a DSTO-AED COE for Damage Mechanics was set up within the CAMT to work mainly on damage modelling, structural health monitoring, and smart materials for sensing and reliability. We took on many of Klaus Friedrich's doctoral students and postdocs to work in the CRC-AS/CRC-ACS and DSTO-AED COE-Damage Mechanics.

4.2.6 *Home to Hong Kong and HKUST*

"The future is not ours to see" has definitely some truth in it. It was a chance encounter with Prof. Pin Tong (董平) in 1991 when I attended his seminar in Hong Kong Polytechnic (HKP) that eventually brought me back home. Tong was just appointed the founding head and professor of mechanical engineering at the newly established Hong Kong University of Science and Technology (HKUST). He had apparently consulted Prof. Max Williams of Pittsburgh (who was Editor of the *International Journal of Fracture*) after we met. Tong had known Williams since the days when he was a PhD student with Prof. Yuan-Cheng Fung (馮元楨) at Caltech in early 1960s. In his own gentle way, Tong was persuasive, we clicked and I accepted to join HKUST. The campus in Clear Way Bay on a mountain overlooking Port Shelter was still under construction in 1991 (Fig. 4.27) and not in use until 1992. It became one of the most beautiful campuses in the world (Fig. 4.28). I was acting head during the period 1994 to 1995 after Tong. The important job for me and Tong was to recruit the best academic staff to the department. To this end, we succeeded with some brilliant appointments. Most came from the US but Prof. Tongxi Yu (余同希) joined us as Reader

Fig. 4.27 With Prof. Pin Tong at the HKUST campus still under construction in 1991.

Fig. 4.28 HKUST campus overlooking Port Shelter in
2021.(https://www.google.com/search?client= firefox-b-d&q=HKUST)

from UMIST. Yu was already a world leader on impact mechanics. HKUST
was still using the British system then and the number of *structural chair*
professors were fixed in each department. There were only 3 in mechanical
engineering already occupied by Tong, Jay-Chung Chen (陳介中) and me.

Fig. 4.29   With my former PhD students. L to R: Limin Zhou, Jang-Kyo Kim (*FHKEng*), me, Yan Li and Jingshen Wu (*ca.* 2012 Beijing Nano).

It is worth mentioning that Prof. Jang-Kyo Kim (retired and now at UNSW) and Prof. Jingshen Wu (now Vice-President of HKUST (Guangzhou)) were appointed Lecturers during this time despite objections from a member of the university's appointment and confirmation committee for the one reason of in-breeding since they were my PhD students at the University of Sydney. Ultimately, Kim and Wu proved themselves as outstanding scholars and capable administrators (Fig. 4.29). I also remember our invitation to Prof. Deborah Chung of University at Buffalo, The State University of New York, to visit us. Unfortunately, this did not eventuate.

An interesting development was the establishment of the *Advanced Engineering Materials Centre* (now *Facility*) with HKD $25 M from Prof. Hsin-Kang Chang (張信剛), Dean of Engineering. This was in response to the *Materials Characterization and Preparation Centre* set up earlier within the School of Science. I became the first Director of the Center. We bought many new processing, testing and characterization facilities for the main research areas at that time on polymer matrix composites, metal matrix composites and polymer blends. The rest is history.

There are many engineering applications (for example, printed circuit boards and surface coatings) in which strong adhesion of thin film on rigid substrate is required. The adhesion energy is usually measured by the blister test under increasing pressure and becomes unstable when a critical pressure

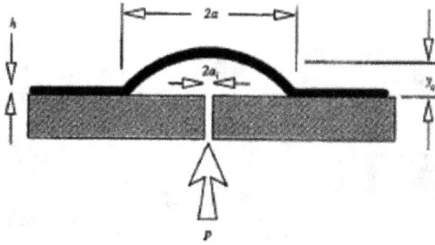

Fig. 4.30   Sketch of a pressurized blister loaded by constant pressure resulting in unstable cracking.

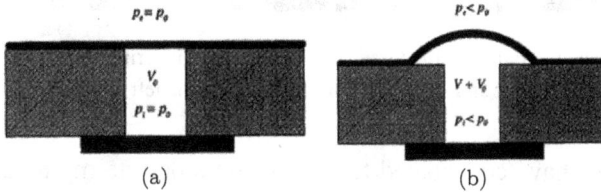

Fig. 4.31   A new blister test with stable crack growth (*Acta Metallurgica et Materialia* 43, 1995, 4109). Pressurized blister with fixed amount of Ideal gas of initial volume V_o and initial pressure p_o when (a) $p_e = p_o$ and (b) $p_e < p_o$.

is reached (Fig. 4.30). In HKUST, Dr. Kai-tak Wan (then a research associate supported by my Australian Research Council (ARC) project grant) and I worked on this film/substrate adhesion problem and we developed a new blister test with stable crack growth. In essence, the method is to allow an isothermal internal expansion of a fixed mass of an ideal gas in a cavity by reducing the external pressure leading to stable crack growth (Fig. 4.31) and measurement of the adhesion energy G_C. We also modified the blister test in which the thin film is loaded instead by a rigid shaft (*Int. J. Fract.* 74, 1996, 181). This also enables a stable interface debonding process to happen and facilitates the adhesion energy G_C to be evaluated. Wan, now a professor at Northeastern University at Boston, has successfully applied these blister techniques and their modified forms to cellular biomechanics in the mechanical characterization and adhesion measurements of cancer cells, bacterial strains, biological tissues, etc.

In just 30 years, HKUST has become a high-ranked university in many league tables and has clearly lived up to its reputation and expectation.

Fig. 4.32   Inauguration of the Graduate School of Engineering (GSE) in 1995 with Pro-Vice Chancellor and Dean John Glastonbury on the left and I am on the right.

I am happy I have contributed to this outcome despite my relatively brief tenure there. In 1995 the University of Sydney set up the first Graduate School in Engineering (GSE) in Australia. I was called and accepted the Director position and the many future challenges of this new graduate school (Fig. 4.32). Before I left HKUST, we tried to get Prof. Tsu-Wei Chou (鄒祖煒) of Delaware University interested in the headship but he declined after visiting the campus with his wife. Chou recently retired after 53 years at Delaware. Finally, Prof. Ping Cheng (鄭平) of the University of Hawaii at Manoa was appointed and he arrived in late 1995. Cheng was elected to the Chinese Academy of Science having moved to Shanghai Jiaotong University after retirement from HKUST in 2002. An outstanding accomplishment of the Department of Mechanical (now Mechanical and Aerospace) Engineering at HKUST is that it has till now produced 3 CAS members (Ping Cheng, Tong-Yi Zhang and Tianshou Zhao) and 1 foreign CAE member (me). This reflects the very high quality of its academics.

4.2.7 *Post-1997 Hong Kong and the City University of Hong Kong*

Life is full of surprises! Things happen when you least expect them. I was on sabbatical in the second half of 1997 as Eminent Visiting Professor in

Materials Science at the Department of Applied Physics and Materials (AP) in City University of Hong Kong (CityU). My wife was with me at the time since we wanted to witness what would be like living in Hong Kong after its handover (or more properly sovereignty returned) to China on July 1, 1997. As we saw it then, nothing really changed and people went on their usual lives. We felt that the *one country – two systems* did work and truly *"horses still run, stocks still sizzle and dancers still dance"*. We were hopeful and optimistic. Our parents were aging and we had thought of spending more time at home. That opportunity finally arrived.

One day in a lift to the Human Resources Office in Cheng Yick Chi Building, I bumped into President Hsin-Kang Chang who was my boss at HKUST when he was Dean of Engineering and I was Acting Head of Mechanical Engineering. He knew then I had decided to return to Sydney. He is a man of taste and particularly French. So, over a beautiful French lunch in Sai Kung, he tried to persuade me to stay. The irony is that he also left HKUST to the University of Pittsburgh sooner than I but was later head-hunted back to CityU. This time he just asked me if I would be interested to join CityU which was only granted university status in 1994. This chance encounter for a few minutes in the lift was a game-changer! A couple years later, Prof. Roderick Wong, then Dean of Science and Engineering, would turn up in Sydney and poached four of our outstanding Sydney engineering colleagues (including David Hill, Hong Yan, Ron SY Hui and me) to join CityU! He was an excellent dean who only hired the best and rapidly built up the high research reputation of this new university (Fig. 4.33). Prof. Stephen Smale, 1966 Field Medal winner on topology in higher dimensions, was a good example. An interesting anecdote I could repeat (and no offence meant) was that Prof. Haydn Chen of UIUC and I were interviewed for the headship of AP and I got it. However, Wong convinced me to take up the headship in Manufacturing Engineering and Engineering Management (MEEM) because he believed I could manage engineering much better. Chen was finally offered AP. I was, in fact, quite ignorant about engineering management at that time. Subsequent exposure and understanding of the teaching and research activities in EM, which is a half of MEEM, removed all my doubts and taught me to respect this apparently "soft" yet could be "hard" discipline.

Fig. 4.33   With CityU of Hong Kong colleagues in Xi'an *ca.* 2001. (L to R): Haydn Chen, Roderick Wong *FRSC*, Stephen Smale *NAS* - 1966 *Field Medallist*, and me.

Fig. 4.34   I was Chair Professor and Head of Manufacturing Engineering and Engineering Management (AMME) at CityU of Hong Kong (2000-02).

When I was the Head of MEEM from 2000 to 2002 (Fig. 4.34), on the teaching side, I introduced the engineering doctorate (EngD) program for manufacturing which required both coursework and a thesis project related to the candidate's workplace. It was an instant success and we had to limit the first intake of students to 9. Later the program was adopted by other engineering disciplines within CityU. This fostered a cohesive bond between industry and the university. On the research side, I strengthened the two applied research labs in the department to help HK industries. One was on advanced surface coatings using physical vapor deposition (PVD) techniques and the other was on smart asset management (for example, structural health monitoring of transport infrastructures, typically, Hong Kong MTR and Chek Lap Kok International Airport). I also collaborated with various colleagues to research on composite coatings, short fiber reinforced polymers and smart materials. In fact, it was during this short 2-years in CityU, I started to work on nanocomposites that would dominate much of my future research in the next two decades.

Superhard nanocomposite coatings for manufacturing industries

During the pre- and post-1997 times, most of Hong Kong's manufacturing industries moved to Dongguan and Shenzhen across the border. However, the highly-skilled mold and die and forming/cutting tools sectors largely remained. Hard surface coatings were required for long service life (e.g., ASM, Johnson Electric) and especially for decoration (e.g., jewellery and watch bracelets). Further, for sustainability and green (or dry) machining without toxic cutting fluids, surperhardness (> 40 GPa) was needed. Unbalanced magnetron sputtering PVD (Teer Coatings Ltd) was preferred due to its lower processing temperature and the metal substrates used. Prof. Edmund Cheung, Prof. Lawrence Li, Prof. Yaogen Shen (my former Research Fellow in Sydney who came to CityU with me and an expert on the physics of thin films) and I obtained substantial combined funding of over HKD $17 M from the Hong Kong Innovation Technology Council (HKITC) and the HK Mould & Die Council. We first investigated the binary nanocomposite system, for example, nc-TiN/a-Si$_3$N$_4$ (where nc is nanocrystalline, and a is amorphous), and examined the microstructures (Fig. 4.35a-b), their Rockwell adhesion, blister failure mode at the interface, cutting performance (measured by number of holes that can be drilled) and wear rate (Fig. 4.35c-d). We then worked on the multilayer systems (e.g., Ti-Si-N, Ti-Al-N, TiN/SiN$_x$, TiN/CN$_x$) which would improve the adhesion and reduce the residual stress between the nanocomposite coating and substrate (Fig. 4.36). Later, Dr Tai Chan, my Victoria Technical School classmate, who was managing the GM-China academic partnership program, provided some financial assistance to the project (Fig. 4.37). In view of CityU's apparent disinterest, GM (US), on its own, filed a patent application which was granted on August 29, 2006. This was

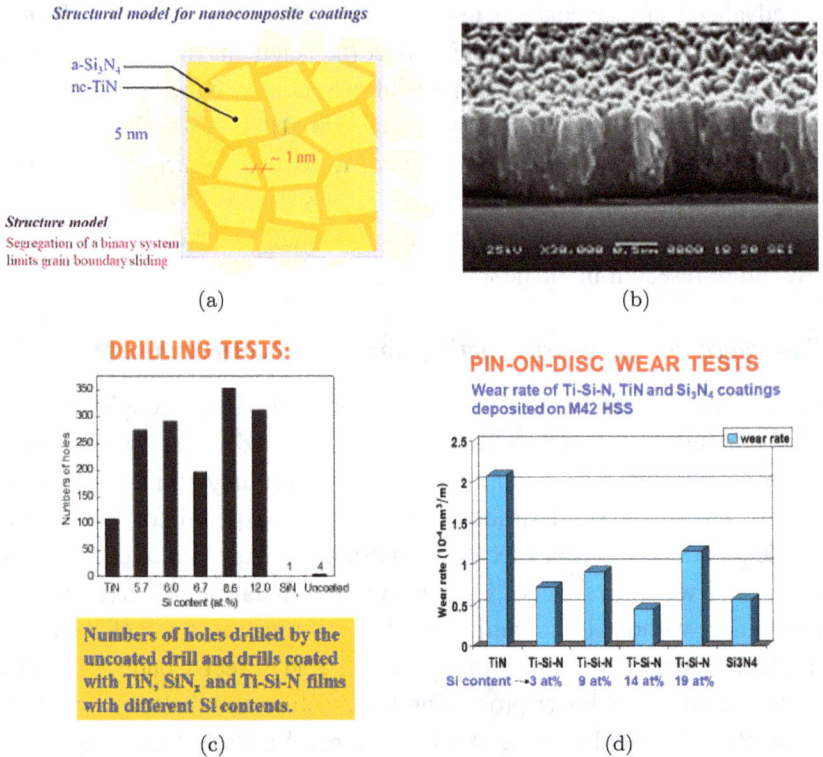

(a)

(b)

(c)

(d)

Fig. 4.35   Superhard binary nc-TiN/a-Si$_3$N$_4$ nanocomposite coatings for cutting tools. (a-b) Structural model and SEM image of nc-TiN/a-Si$_3$N$_4$. (c-d) Drilling and wear tests.

the US Patent #7,097,922 on *"Multi-layered Superhard Nanocomposite Coatings"* and I was one of the inventors.

Short-fibre reinforced polymers

I continued my research on short-fibre reinforced polymers (SFRPs) with Prof. Sie Chin Tjong and Prof. Robert Kwok-Yiu Li from AP. SFRPs are a class of composites of great practical importance. They have been widely used in automobiles, sporting goods, dental, electronic and electrical industries due to their low-cost and easy processing. In late 1990s to early 2000s in Hong Kong, plastics waste and plastics recycling were hot topics. We were particularly interested in polyethylene terephthalate (PET) which was a widely used consumer plastic. Our major effort was on *value-added recycling*. To achieve this objective, we could incorporate functionalised

Layered structure $(Ti_{0.58}Al_{0.42}N)$ on steel and tool insert by PVD

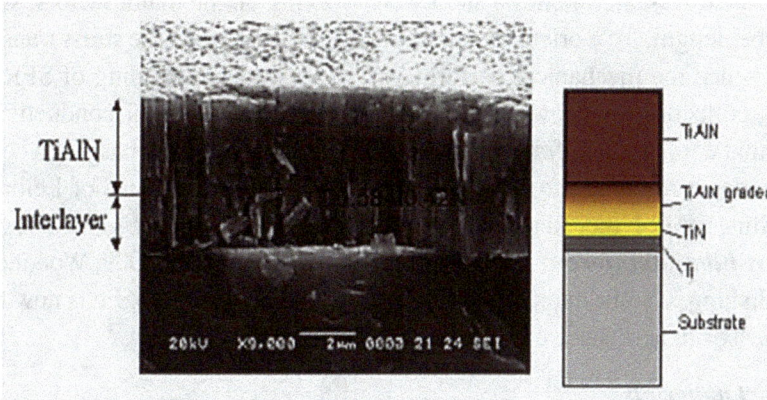

Fig. 4.36   Superhard multi-layered nanocomposites coatings for cutting tools and inserts.

Fig. 4.37   With my VTS classmate Tai Chan (R), GM-China, in CityU's advanced coating laboratory.

rubber particles to toughen the recycled PET, but the losses in strength and stiffness had to be compensated by adding short glass fibers. We did the same for other polymers such as PP and PA6. We studied the essential work of fracture (EWF) method for these binary and ternary systems. We determined and compared their specific EWF (or fracture energy) values in terms of their specific deformation and toughening mechanisms. At the same time, Dr. Shaoyun Fu (who followed me from Sydney to CityU and appointed Research Fellow) and I were working on the major factors, such as fiber length, fiber orientation and anisotropy, that affect the stress transfer and hence the mechanical performance in tension and bending of SFRPs. This collection of knowledge and later works on electrical conductivity, thermal conductivity, fracture, creep and fatigue formed the basis of a comprehensive research monograph that we and Dr. Bernd Lauke of Leibnitz Institute for Polymer Research Dresden wrote: *Science and Engineering of Short Fibre Reinforced Polymer Composites* (1st edition, 2009, Woodhead Publishing, Cambridge; 2nd edition 2019, Elsevier, London). Fu is now full professor in aerospace engineering at Chongqing University.

Smart materials

My interest on smart materials started with the establishment of the DSTO-AED COE on Damage Mechanics mentioned in Section 4.2.5.1 above. Such smart materials include functionally graded materials (FGMs), piezo-electric, piezomagnetic and electromagnetic materials and any combination thereof. They are embedded in matrices to make devices and our main focus has been on their service integrity and reliability under coupled mechanical (or thermal), electrical and magnetic loadings. Dr. Qinghua Qin (an ARC QEII fellow and later professorial fellow in the CAMT; and finally a full professor at the Australian National University until his recent retirement) had already done important work on the effective properties and crack problems in piezoelectric materials. This was continued by Dr. Baoling Wang who came to CityU from Shizuoka University as Research Fellow focusing on the fracture mechanics of FGMs and piezoelectric/piezomagnetic composite materials. The effects of the crack-face electromagnetic boundary conditions were studied. Later in 2013, Wang joined the CAMT on a 5-year research fellowship on smart materials funded by the ARC.

4.2.7.1 *To stay or not to stay*

After about a year, my wife, Louisa, and I were quite settled and we lived on campus. Importantly, we could spend more time with our parents and immediate family members. We thought we might just stay in Hong Kong. I was also happy with working in CityU despite the daily routine and chores as the head of MEEM, and the research was progressing well. I had just been awarded a prestigious group research project of HKD $4.0 M on *"Polymer Nanocomposites"* in collaboration with Prof. Chi-Ming Chan and also Prof. Jingshen Wu of HKUST and Prof Limin Zhou of Hong Kong Polytechnic University (PolyU was upgraded from HKP in late 1994) by the Hong Kong Research Grants Council. We would investigate the processing methods, control of microstructures, mechanical properties, failure and deformation mechanisms of polymer nanocomposites which contain nanofillers (e.g., calcium carbonate and clay) in polymer matrices (thermosets and thermoplastics). I should add that, in early 2001, I was pursued by Sydney University's Research Office to apply for a Federation Fellowship instigated by the Australian Government as part of its Backing Australia's Ability initiative to lure back high-profile Australian researchers from foreign institutions. I submitted a project: *"Deformation and Fracture Studies on Polymer Nano-Composites"* and never thought I would succeed due to the extremely tough competition worldwide. I had forgotten about this matter until one summer day in 2001 I received an unexpected overseas phone call from the CEO of the ARC that I was selected as one of 15 successful candidates. The fellowship was for 5 years with an A$1.25 M research support and an attractive salary. It is a research-only position with no administrative duties.

So, to stay in Hong Kong or to return to Sydney, that was the question (Fig. 4.38– *New Scientist, 27 October 2001, p. 57*). In the end, my stronger desire to build a new research team to work on the engineering and science of an emerging class of materials - *polymer nanocomposites* - made my decision easier. However, I was very much saddened to leave the department I led for over 2 years and all the colleagues with whom I had built a strong rapport and mutual respect (Fig. 4.39). I tried to interest Prof. Andrew Y.C.

Fig. 4.38   Wither Hong Kong or Sydney?

Fig. 4.39   With MEEM academic staffs before my return to Sydney in 2002. I am standing 5th from right in the front row.

Nee (倪亦靖) of the National University of Singapore (NUS) to take over the headship of MEEM. He would be the perfect candidate. I knew he was the younger brother of the novelist/screenwriter Ni Kuang (倪匡) and the novelist Yi Shu (亦舒). But it is only recently did I know that he graduated from Kowloon Technical School which was founded as a sister school of Victoria Technical School. Nee and his wife visited the CityU campus but decided in the end to remain in Singapore. He had retired from NUS in 2018. Nonetheless, MEEM continued to flourish, and for a short time, Prof. Steve Hsu of tribology fame at NIST in Gaithersburg, Maryland, was the head of the department. I am very pleased to learn that today MEEM has evolved into 3 separate engineering departments of Mechanical, Advanced Design and Systems, and Biomedical.

4.2.8 *Return to the University of Sydney and new challenges*

Since returning to Sydney from Hong Kong on an inaugural Australian Federation Fellowship in October 2002, and in the ensuing years, my major research challenges were on polymer nanocomposites which I had already gained some knowledge in CityU. Besides the original proposed studies on deformation and fracture, I included key allied research areas on the chemistry of materials (to understand the complex processing of these nano-materials through *in-situ* polymerisation, melt compounding and solution blending) and multifunctional properties (to identify their engineering applications). By definition, a polymer nanocomposite contains fillers with at least one dimension less than 100 nm in a thermoplastic or thermoset matrix.

The work on the fracture mechanics of smart materials was extended with Dr. Baolin Wang who became an ARC Research Fellow at the CAMT (now an associate professor at Western Sydney University) and a former postdoc, Dr. Xiaohong Chen of Goodrich Structures (USA). Chen and I co-authored a research monograph on *Fracture Mechanics of Electromagnetic Materials: nonlinear field theory and applications* (Imperial College Press, 2013). It was gratifying that, at about the same time, a former PhD student, Dr. Ee-Hua Wong of Canterbury University in New Zealand, and I finally finished our book on *Robust Design of Microelectronics Assemblies against Mechanical Shock, Temperature and Moisture* (Woodhead Publishing/Elsevier, 2015), which is critical to understand the mechanics and hence safeguard the reliability of parts and components of packages and assemblies of modern microelectronic devices.

I also embarked on several new research areas, some were finished and some are still ongoing today. Typical topics include: (a) nanostructure toughening of epoxies with reactive block copolymers or block ionomers with and without nanofillers (Prof. Qipeng Guo of Deakin University); (b) size-dependent mechanical properties of electrospun nanofibers (Prof. Josh Shing-Chung Wong formerly of University of Akron and now with LyondellBasell), and (c) epoxy-based nanocomposite as electronic packaging underfill materials for easy processing, efficient heat dissipation and electrical insulation (Prof. Xiaolin Xie of Huazhong University of Science and Technology (HUST)).

Deformation and fracture of polymer nanocomposites

(a) In mid-1980s, I visited Japan for 6 weeks supported by a Japan Society for the Promotion of Science (JSPS) fellowship, and my research then was mainly on cementitious composites and toughened ceramics. I remember, in the Toyota Research Laboratories in Nagoya, I met Dr. Akane Okada who showed me their work on organoclay/nylon 6 composites for under-the-hood automotive parts. The tensile strength,

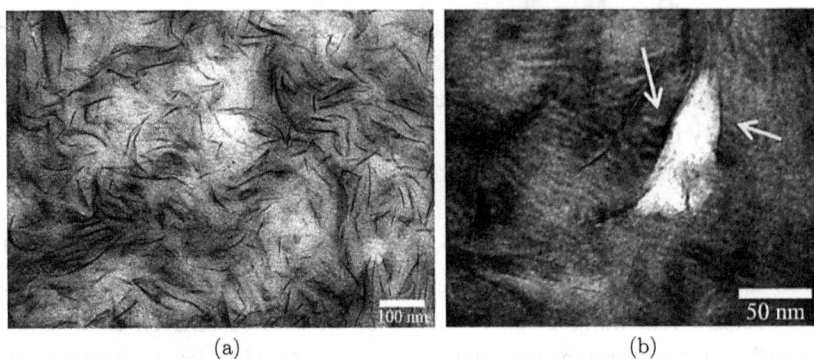

(a)   (b)

Fig. 4.40   TEM micrographs of organoclay/PA 6 nanocomposite (90/10). (a) Unloaded
showing clay layers in PA 6 matrix. (b) Transcrystalline layers of PA6 (identified as white
layers) remain tethered to the delaminated clay forming a void.

elastic modulus and heat distortion temperature were significantly
increased but the elongation to break was much reduced which implied
that the fracture energy would be dramatically decreased. But why?
We later showed convincingly with TEM micrographs that when a
clay/nylon sample was progressively loaded (*Macromolecules* 40, 2007,
123), the nylon transcrystalline lamellae regions originally organized
transverse to the exfoliated or intercalated rigid clay layers remained
intact and did not activate any major matrix deformation processes,
even after delamination of the clay layers (Fig. 4.40). Hence, large
improvement in fracture energy could not be achieved.

(b) A basic question follows. In many thermoplastic matrix nanocomposites, interface debonding is the precursor which triggers other deformation processes to provide a high fracture toughness. So, is there a
critical size requirement of the nanofillers above which there is effective
toughening and below which there is none? We have examined this
issue using meso-mechanics (*Compos. Sci. Technol.* 70, 2010, 861) and
demonstrated that this critical size depends on the absolute value of the
work of adhesion. In the adhesion range of 0.01-0.40 J/m^2, depending
on the elastic modulus ratio of nanofiller to matrix, the critical radius of
a spherical particle can be tens to several hundreds of nanometers; and
for a penny-shaped particle with loading parallel to its faces, the critical
particle size is at least several micrometers for an aspect ratio of 100.

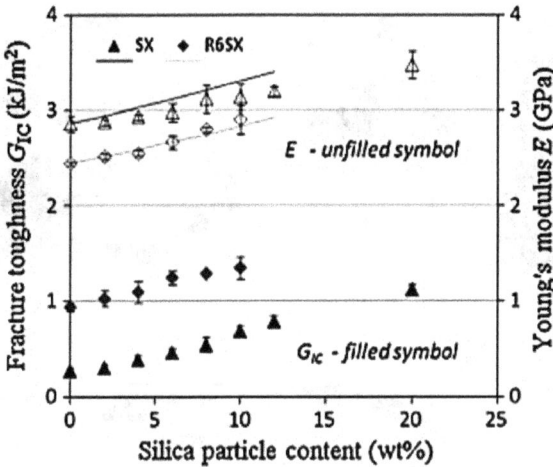

Fig. 4.41   Fracture energy and Young's modulus values of epoxy-based nanocomposites. R: rubber, S: silica, and RS: hybrid rubber/silica; X; wt.%. Here, R is fixed at 6 wt.% and S varies from 0 to 20 wt.%.

(c) There are clear benefits of incorporating nanofillers such as core-shell rubber (~100 nm) and silica (20-30 nm) into thermoset epoxy matrices to increase their fracture energy (and elastic modulus) values (Fig. 4.41). Incorporation of hybrid nanorubber and nanosilica in epoxy gives only an additive effect on the fracture energies.

Composite laminates toughened by hierarchical fibres, interlaminar inter-leafs and 3D foams

Composite laminates are made up of many individual plies to construct structural parts and components like the fuselage and wing sections of air-crafts. A major weakness is the delamination between the plies which may be caused by external loading (for example, impacts due to hail stones during flying and/or hammer drops on maintenance inspection) and extreme service temperature fluctuations. Beams containing a long edge crack in the mid-plane can be conveniently used to measure the delamination resistance due to opening of the upper and lower halves or relative sliding of these under 3-point bending (Fig. 4.42a-b). In Section 4.2.4, we mentioned that perforated or full plastic thin films could be used to enhance the inter-ply fracture energy against delamination cracks. In view of recent advancements

(a)   (b)

(c)

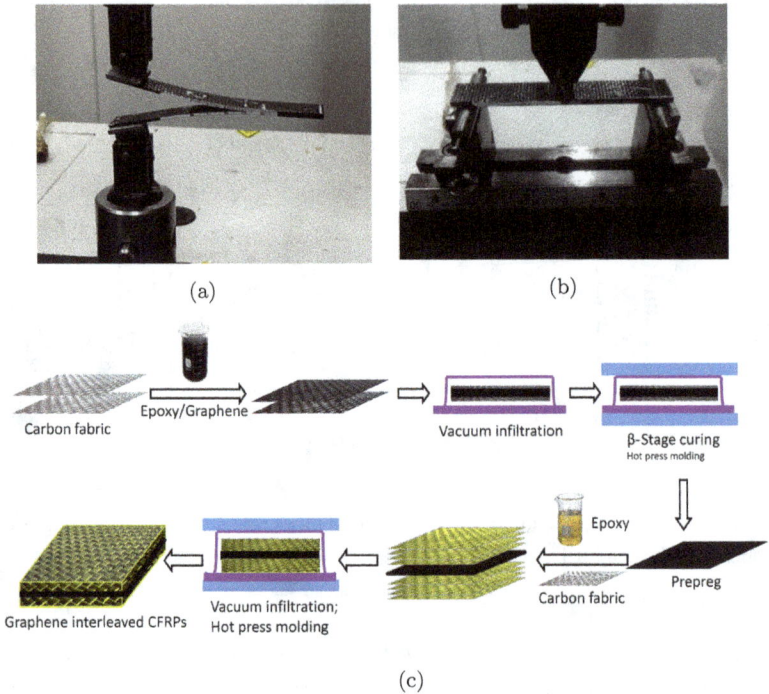

Fig. 4.42 Measurement of delamination fracture energy with beams containing a mid-plane edge crack under (a) opening mode and (b) relative sliding mode of upper and lower halves. (c) General preparation process of interleaf prepreg and interleaved composite laminates.

in nanotechnology, we can adopt these new methods to engineer the interfaces with improved mechanical properties.

(a) The first method is to prepare toughened interleafs which are inserted between the plies of the composite laminate (Fig. 4.42c). We have worked on many different types of interleafs for CF/epoxy composites with generally positive outcomes. These include thermoplastic (polysulfone, nylon, etc.) and thermoset (epoxy resin) interleafs containing nano-, and sometimes micro-, fillers (CNTs, cellulose nanocrystals, short carbon fibres, etc.). Of particular note is our work on 3D interconnected graphene foam/epoxy nanocomposites as interleafs in glass fibre/epoxy composite laminates which display quite large fracture energies against opening and sliding delamination as well as substantially improved interlaminar shear strength (ILSS) (Fig. 4.43).

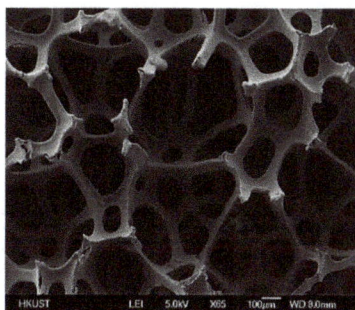

(a)

(b)

(c)

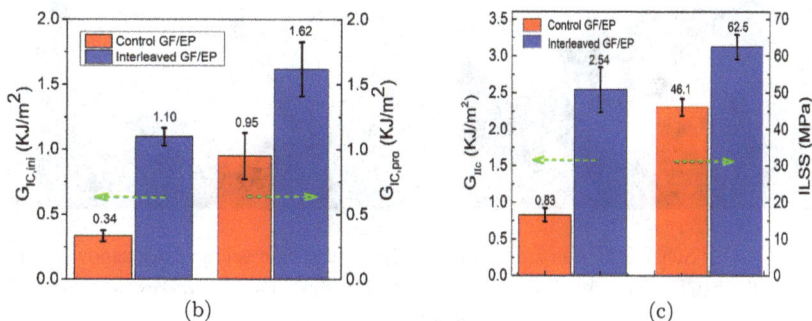

Fig. 4.43   Glass fiber/epoxy interleaved with 3D graphene foam showing (a) microstructure of foam, and (b) improved opening fracture energy (G_{IC}) at crack initiation and propagation, and (c) sliding fracture energy (G_{IIC}) and interlaminar shear strength (ILSS).

(b) A second method is to use hierarchical fibers such as by growing zinc oxide (ZnO) nanowires (or CNTs) onto carbon fibers or carbon fiber fabrics with a hydrothermal method (or low temperature flame synthesis). An AFM setup then measures their bond strength (Fig. 4.44a). Surface treatment of ZnO by polydopamine (PDA) is found to increase the bond strength between ZnO and carbon fiber (Fig. 4.44b-d). Subsequent work with ZnO (or CNT) on carbon fiber fabrics also demonstrate highly increased opening and sliding delamination fracture energies.

Multifunctional properties of polymer nanocomposites

(a) The research frontiers on polymer nanocomposites have shifted in the past decade from purely mechanical properties to their multifunctional properties. Our early efforts have been summarised in a monograph

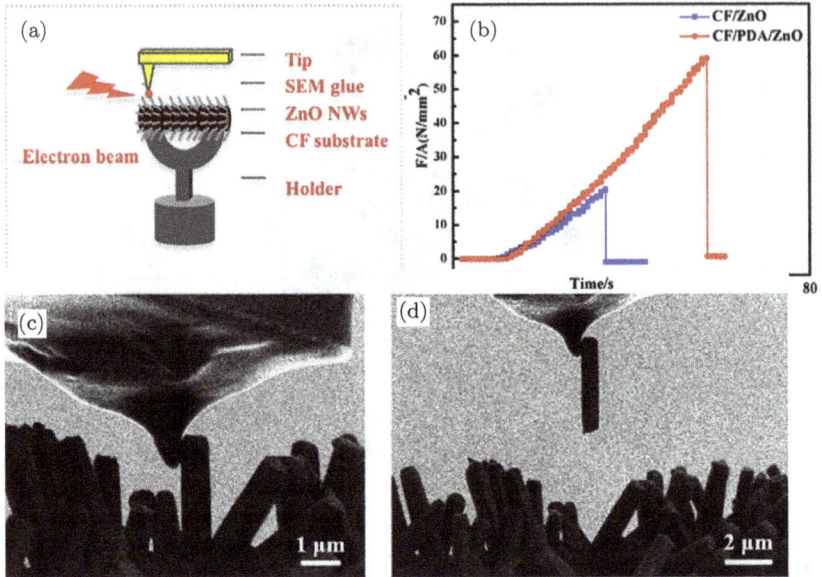

Fig. 4.44   (a) Schematic of *in-situ* pulling off individual zinc oxide (ZnO) nanowrire from carbon fiber (CF) using an AFM tip. (b) Pull-off force per unit area *versus* time during the pull-off process showing polydopamine PDA enhances ZnO/CF bonding. (c-d) SEM images of pull-off of a ZnO nanowire from CF surface.

on *Polymer Nanocomposites – towards multi-functionality* (Springer-Verlag, London, 2016) which I co-authored with former CAMT alumni, Prof. Aravind Dasari (now at Nanyang Technological University, NTU) and Prof. Zhong-Zhen Yu (now at Beijing University of Chemical Technology, BUCT). There are too many functional properties, singly or conjointly, which can be designed and achieved with polymer nanocomposites for specific applications. We have worked on some of these which include, for example, friction/wear properties (*MSER*, 63, 2009, 31), fire retardancy (*Prog. Polym. Sci.* 38, 2013 1357), permeability, optical transparency, thermal and electrical conductivities, EMI shielding, dielectric properties, etc.

(b) Permeability is an important barrier property that has important applications. For example, Wilson double core tennis balls, Coors Light plastic beer bottles (which would be phased out soon but not on technical grounds), etc. We showed that by incorporating very low loading of

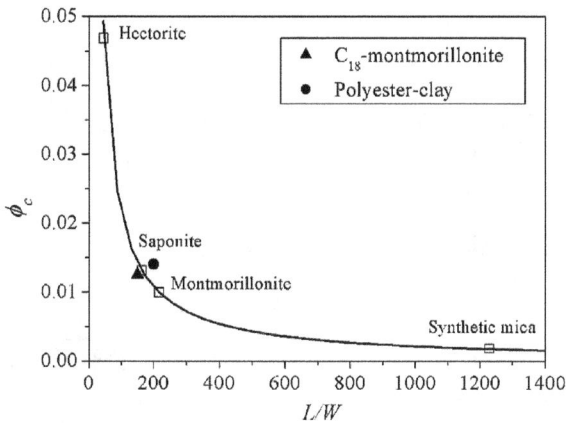

Fig. 4.45 Theoretical curve for critical volume fraction ϕ_c *versus* aspect ratio L/W of silicate platelets with 3D randomness orientation. Experimental values for several typical clay fillers are shown (open squares). Experimental data of O_2 gas permeability in polyester-clay (solid circle) and C_{18}-montmorillonite (solid triangle) are also given.

exfoliated 2D silicate fillers into polymer matrices, the permeability of polymer-layered silicate nanocomposites decreases remarkably. With Prof. Chunsheng Lu of Curtin University, we proposed an improved tortuosity-based model, in which the pathway of a diffusing gas/liquid molecule in the exfoliated nanocomposites is just like a self-avoiding random walk. Using the concept of group renormalization, we calculated the critical volume fractions (ϕ_c) of clay fillers with different aspect ratio (L/W) for minimum permeability, which were in good agreement with available experimental data (Fig. 4.45).

(c) A high transmittance is required for optical functional materials for successful practical applications, such as optical lenses, LCDs, optoelectronic packaging, etc. The introduction of inorganic nanoparticles, even at low contents, into transparent polymers increases their mechanical properties. However, this often leads to opaque nanocomposites due to light scattering by the nanoparticles because of the refractive index (RI) mismatch between nanoparticles and polymer matrices. So, RI matching is the best strategy. In collaboration with Prof. Shaoyun Fu of Chongqing University, we synthesized highly transparent epoxy (EP-400) nanocomposites by adding core-shell structured silica-titania (S-T) nanoparticles (1 wt.%) where the titania coating content could be

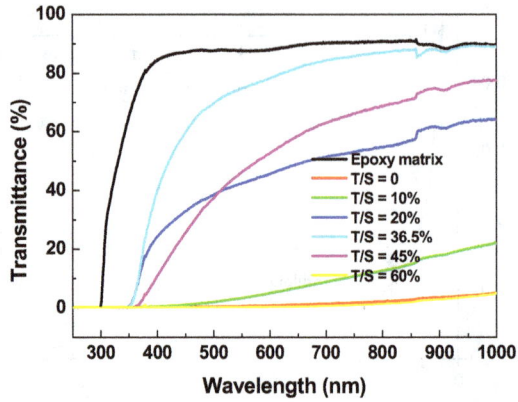

(a)

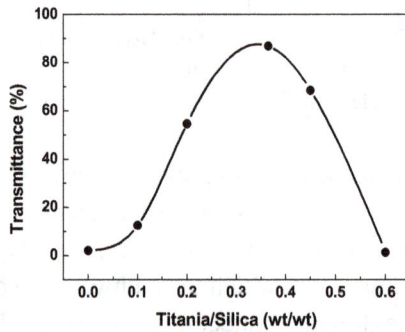

(b)

(c)

Fig. 4.46   Effect of TiO_2 content on optical transparency of epoxy containing 1 wt.% of core (SiO_2)–shell (TiO_2) nanocomposites: (a) with wavelength and (b) at 800 nm wavelength. (c) At 36.5 wt.% TiO_2 optical transparency is best.

varied and the RI changed to match that of the epoxy matrix. In this way, we showed that optimal transmittance of the nanocomposite could be obtained at a titania shell weight content of 36.5% (Fig. 4.46). The Si diameter was ~59 nm and Ti coating thickness was ~5 nm. This strategy was also proven for 2 and 10 wt.% of S-T nanoparticles in EP-400 with no issues of dispersion difficulties.

Multifunctional electrospun nanocomposite fibers

My interest in electrospinning started in the late 2010s when I interacted with Prof. Josh Wong of Akron University and Prof. Chen-Chi M. Ma of National Tsing Hua University. Initially, Wong and I were interested in the processing-microstructure-mechanical properties relationship of electrospun neat fibers, especially their size effect (*Compos. Sci. Technol.* 70, 2010, 703) on strength and stiffness. When I recruited his doctoral student, Dr. Avinash Baji, on my Federation Fellowship project, we refocused on the functional properties and applications of electrospun fibers. Thus, we investigated electrospun PCL/PLA fibers for scaffolds and dressings in tissue engineering, electrospun fiber reinforced composites, electronspun interleafs, filtration membranes, and composite nanofibers containing different filler types, e.g., CNT, organoclay, HAP, silica, magnetic particles in PA6, PVDF, PU, etc. matrices for different functions. Prof. Ma sent his PhD students to work with us and supported by the Taiwan National Science Council aiming to manufacture flexible transparent conductive films (TCFs) for modern electronic devices. In one example, we used (0.05 wt.%) graphene nanosheets (GNS) decorated with silver (Ag) nanoparticles which were self-assembled onto the surfaces of PU nanofibers. Besides the good bending durability, the results displayed surface resistance of 150 Ω/sq and 85% light transmittance comparable to the brittle ITO with $\sim$ 100 Ω/sq and 85% transmittance (*Carbon* 50, 2012, 3473).

4.2.9 Connections to PolyU and my beloved Hong Kong

4.2.9.1 Research collaborations on electrospinning with PolyU

As mentioned in Sections 4.2.1 and 4.2.2, I grew up in Hung Hom and was emotionally attached to it. I knew the neighborhood very well, especially near the site of the old Hong Kong Technical College (more fondly called TC) where now stood the much larger campus of HKP and subsequently PolyU. Interestingly, the Junior Technical School, which was the predecessor of my high school, VTS, shared with TC the same Red Brick House in

Wan Chai from 1947 to 1957. Many of my VTS classmates went through TC to finally become chartered surveyors and accountants, since these two disciplines were not offered by the local universities and were in high demand, then and now, in Hong Kong. I had a long and strong association with TC, then HKP and then PolyU over the last few decades. I taught the ordinary certificate (OC) subject on *strength of materials* to mechanical engineering evening students at different locations when I was still working on my doctoral thesis project at the University of Hong Kong in early 1970s. Years later, I served in mid-1990s to early 2000s in various capacities as external examiner, industry committee member, department academic advisor and department review committee external member for PolyU's department of mechanical engineering (ME) and applied physics (AP). But there were no research collaborations.

However, yet another chance encounter on campus sometime in 2007 with Prof. Jan-Ming Ko, who was senior vice-president of PolyU at the time, changed all this. Ko knew me from a long way back when we were both research students in HKU. He came from civil engineering and I from mechanical engineering. With the full support of Prof. Jian Lu (who was then Head of Mechanical Engineering at PolyU and now Chair Professor in CityU), Ko offered me a part-time visiting chair professor position and provided substantial funding of ~HKD $2 M for a 5-year research program on *Electrospinning of 1D carbon-based nanofibres as electrodes for Li-ion batteries* from 2008/09 to 2012/13. This was a new research direction for me and for Prof. Limin Zhou who would supervise the projects within the program and managed the allocated funding as a full-time professor. Zhou was my former PhD student in mid-1990s at Sydney and had made many original contributions on the mechanics of composite fibre-matrix interfaces. We recruited a PhD student, Yuming Chen, who had thought of going to Sydney for doctoral studies with me. We later enrolled his wife, Xiaoyan Li, as another PhD student to the research team for this program which also included Prof. Haitao Huang from PolyU's AP department. Chen spent 6 months with Prof. John Goodenough at UT-Austin to work on high performance electrodes for Li-ion batteries. We are honored to have co-authored a few papers with Goodenough, who won the 2019 Nobel Prize in Chemistry for his pioneering Li-ion battery work. Chen and Li are now full

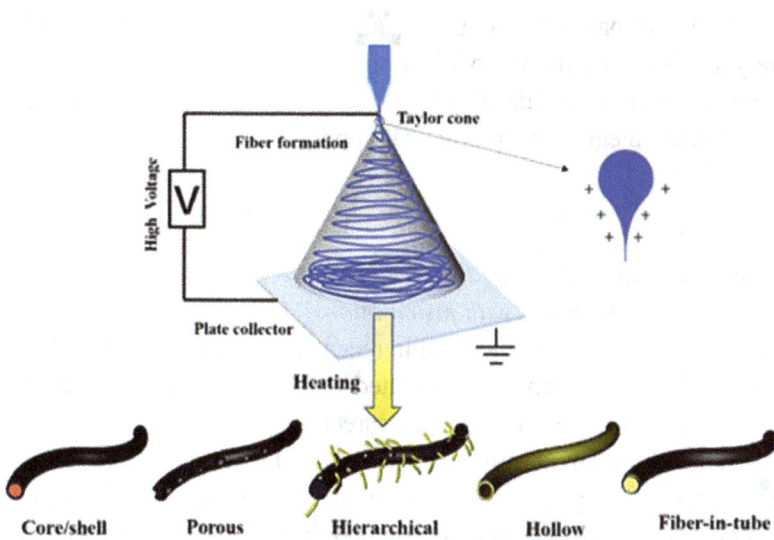

Fig. 4.47   Schematic illustrations of structural design through electrospinning engineering. (Courtesy Yuming Chen)

professors in their *alma mater*, Fujian Normal University, leading research projects on energy storage materials and devices.

Electrospinning is a versatile approach to construct various structural forms of the carbon nanofibers from core/shell, porous, hierarchical and hollow to fiber-in-tube (Fig. 4.47). They can be collected on a flat plate, a rotating drum or two parallel plates. After calcination treatment, they can be used to make electrodes. Through these designed fiber forms and architecture, we can improve their electrochemical performances. For example, a porous structure enables efficient transport and diffusion of the electrolytes and ions. Porous and/or hollow structures can also mitigate the structural strain and buffer volume changes of the electrode materials during repeated charging/discharging, thus increasing the cyclic stability.

There are other examples with other structural forms such as fiber-in-tube and hierarchical fiber. We have recently reviewed the electrospinning strategies for battery materials (*Adv. Energy Mater.* 11, 2021, 2000845) in general, and 1D nanostructured electrodes for energy storage (*Energy Storage Materials* 5, 2016, 58) and sodium-ion batteries (*Adv. Fiber Mater.*

4, 2022, 43) in particular. With the increasing demand of more efficient battery anodes and cathodes for electronic devices, I believe electrospinning research in this field will develop fast and new science will be achieved even faster than its integration for practical applications.

4.2.9.2 *Hong Kong: intertwined emotions and memories*

My memories severed not by distance and time
Forever they are threaded with my emotions

    Hong Kong may have changed in recent years, but it is still my beloved city in which I grew up, was educated and which had been the spring-board for me to build an academic career in Australia through sojourns in the US and the UK. I still have many fond memories of and emotional attachments to this place, its people, its history and traditions, and its old landmarks which are permanently impressed in the depth of my heart. And I still have strong loyalties and feelings for the University of Hong Kong where I spent almost 8 years as an undergraduate and then a postgraduate (1966-73) in the old engineering buildings which had long given way to the concrete flyovers erected on Pokfulam Road. During the past 30 years, I had worked in different capacities over different times in HKUST (1993-95), CityU (2000-02), PolyU (2008-13), and HKU (2014-18). They made up different chapters tightly interwoven into the fabric of my life story. I had never thought of working in Hong Kong since I left in 1973. The accidental meeting with Prof. Pin Tong in 1991 and his invitation to join HKUST reset my destiny. This first opportunity truly set in train my later affiliations with CityU, PolyU and HKU. I believe I have left my marks in these institutions. Whatever they are, with passing time, many will fade but a few will remain. I was therefore very pleased for the invitations to return to campus to celebrate the 20th (2011) and 25th (2016) anniversaries (Figs. 4.48 and 4.49) of HKUST and the department of mechanical and aerospace engineering I helped establish. Same invitations also came from my alma mater, HKU, for the celebrations of the 90th (2001) and 100th (2011) anniversaries of mechanical engineering, and from CityU for the 30th (2014) anniversary of MEEM.

    Broken lotus roots bridged by fibrils are not truly separated. Today, I still have good continuing Hong Kong connections through my fellowship

Fig. 4.48 Celebrating 20th Anniversary of HKUST with ME academic staff in 2011. Sitting from left: 5th (me), 6th (Ping Cheng, *CAS Academician*), 7th (Pin Tong), 8th (Tongxi Yu) and 9th (Ching-Ping Wong, *NAE*). Standing from right: 6th (Tianshou Zhao, *CAS Academician*), 8th (Tongyi Zhang, *CAS Academician*) and 10th (Jang-Kyo Kim *FHKEng*).

Fig. 4.49 Celebrating 25th Anniversary of HKUST with ME colleagues in 2016. I am standing on the first step of the stairs in black suit from the right. Pin Tong is behind me.

with the Hong Kong Academy of Engineering Sciences. More importantly, I still have passionate emotions for my beloved Hong Kong and deep feelings for my siblings and their families (Fig. 4.50). My mother sadly died at age 101 on May 28, 2021. Due to the pandemic, I could not return to Hong Kong for her funeral. This regret will forever be with me.

(a)

(b)

(c)

Fig. 4.50　(a) (L to R) 2015 my mother's 95th birthday with Chung-Ho, Chung-Sum and Yiu-Shing in back row, Yiu-Shang and me seated in front row; and (b) extended family members. I am seated first on the right. (c) (L to R): 2020 my mother's 100th birthday with Chung-Ho, Yiu-Shing and Chung-Sum standing, me and Yiu-Shang seated in front row.

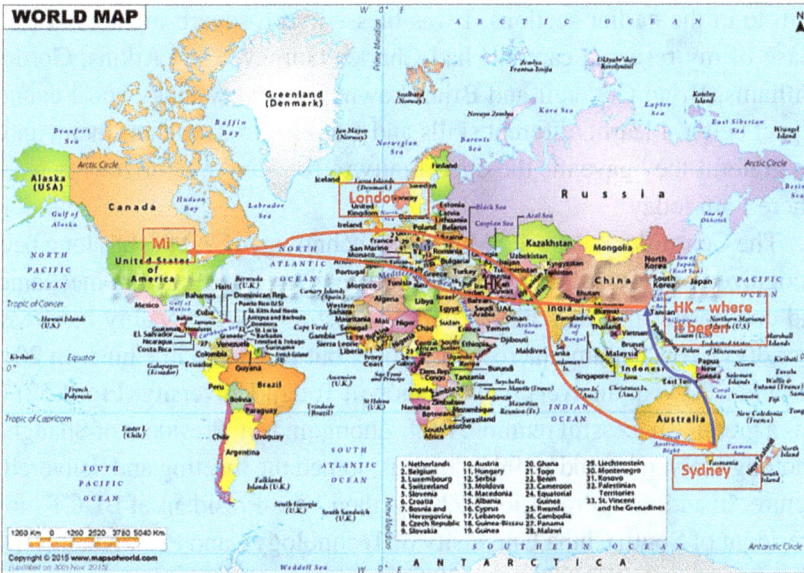

Fig. 4.51 In pursuit of an academic and research career, I travelled East, West and South.

4.2.10 *Looking back on 50 years of research*

Over the past five decades, from where it began in Hong Kong, I travelled West, I travelled East and I travelled South to pursue an exciting academic and research career (Fig. 4.51). I have not been disappointed, I achieved more than I expected and I do not have any regrets. A motto I have always treasured is: *"Happiness scales inversely with expectations"*. Hence, I only set the level of expectation that I can achieve each time. Then I am always happy. Another motto for me is: *"We all have a use-by date – some come early some come late"*. I know the curtain will be drawn one day and I must leave the center stage. But until that day, to quote Jack London, *"I shall not waste my days and I shall use my time"* to climb maybe a few more heights in the *Mountains of Knowledge*.

4.2.10.1 *My mentors and the CAMT alumni*

In academia as in life, it is very important to have excellent mentors who will counsel you with wise advice at each stage of your career development to reach your peak potential. You may be lucky but you may be not. As

I wrote in the earlier sections, I was blessed with superb mentors at each phase of my research career. I had Charles Gurney, Tony Atkins, Gordon Williams, Brian Cotterell and Brian Lawn; I was very lucky and I cannot expect better. I learnt different skills and qualities from each of them; and I am grateful they gave me the opportunity to broaden my horizons to reach where I am today.

The Centre for Advanced Materials Technology (CAMT) has long been recognized as the cradle for producing some of the finest fracture mechanics and composite materials researchers in Australia. We probably have well over 300 CAMT alumni across the globe, but many are in China. In 2019 we held a 30-year anniversary workshop in Tongji University (Fig. 4.52). It was a greatly successful reunion. Prof. Zhongqin Lin, President of Shanghai Jiaotong University and a 1994 alumnus, joined the meeting and delivered a lecture. In addition, Prof. Liqun Zhang, then Vice-President of BUCT (now President of South China University of Technology), and Prof. Xiaolin Xie, Vice-President of HUST and a 2003 alumnus, also presented lectures on their latest research (Fig. 4.53).

I am very grateful that the CAMT organized in 2006 a symposium for my 60th birthday at the University of Sydney (Fig. 4.54-4.58) and in 2016 another symposium for my 70th birthday in HKUST (Fig. 4.59-4.61). Special issues in *Compos Sci and Technol* (67, 2007, 149-324) and in *Eng. Fract. Mech.* (74, 2007, 1007-1202) contain papers in honor of my 60th birthday. Prof. Chun-Hway Hsueh (at Oak Ridge National Laboratory) could not come in 2006 but he sent a sketch of my portrait, later framed for conference guests and participants to sign on (Fig. 4.54). He is such an accomplished artist other than a remarkable materials scientist!

I specially thank Prof. Victor Li of the University of Michigan and Prof. Klaus Friedrich of the University of Kaiserslautern who came and lectured on both occasions. Friedrich gave a thorough description of my research activities in his 2016 HKUST opening lecture (Fig. 4.60). I must also mention that Friedrich attended the 2019 CAMT 30-year anniversary workshop in Tongji University. I am equally thankful to Prof. Huajian Gao of Brown University (now at Nanyang Technological University), who also presented a lecture at my 70th birthday symposium at HKUST. Due to the pandemic situation in China, my 75th birthday symposium was to be held in Wuxi, but kept on postponing. Eventually, it took the form of 6 webinars

(a)

(b)

(c)

Fig. 4.52   CAMT group photos. (a) 1995, I am standing 2nd from left. (b) 2003, I am 5th from the left on the second row. (c) 2019 – 30th anniversary of CAMT held in Tongji University, Shanghai, sitting row from left – 4th (Xing Ji), 6th (Xiaolin Xie, VP HUST), 7th (me), 8th (Klaus Friedrich), 9th (Lin Ye *FTSE*); and from right – 2nd (Liqun Zhang *CAE*, President SCUT), 6th (Yan Li), 7th (Limin Zhou) and 8th (Tin-You Fan).

Fig. 4.53   Invited speakers at 2019 CAMT Tongji Symposium. (a) President Zhongqin Lin *CAE* (SJTU). (b) President Liqun Zhang *CAE* (SCUT). (c) Prof. Xing Ji (Tongji) (d) Vice-President Xiaolin Xie (HUST).

Fig. 4.54   My portrait sketched by Chun-Hway Hsueh (formerly of Oak Ridge National Laboratory, now at National Taiwan University) for my 60th birthday (2006).

Fig. 4.55 Symposium held at Sydney University of my 60th birthday (2006). Front row (L to R): Brian Lawn *NAE*, Gordon Williams *FRS*, me, Brian Cotterell, Victor Li and Mark Hoffman *FTSE*. Second row (R to L): 2nd (Xiqiao Feng), 3rd (Tongyi Zhang *CAS*), 5th (Debes Bhattacharyya *FRSNZ*) and 6th (Klaus Friedrich). Toshio Tanimoto (President of Shonan Institute of Technology) in suit and tie is behind Bhattacharyya.

Fig. 4.56 CAMT alumni at my 60th birthday symposium. Front row: I am 2nd from left and Lin Ye *FTSE* 2nd from right. Back row: Mark Hoffman *FTSE* 2nd from right.

Fig. 4.57   With Jie Tong (L) of Portsmouth, me and Yan Li (R) of Tongji.

Fig. 4.58   With Liangchi Zhang *FTSE* (L), me, my wife and Victor Li (R) of UM.

organized by the journal, *Nano Materials Science*, which also published a Special Issue in my honor: *Trends in Nanomaterials and Nanocomposites: Fundamentals, Modelling and Applications – Part A and Part B*, (Guest editors: LC Tang and SY Fu), 2022, Vol. 4, Issues 2 & 3, pp 61-294. These lectures were delivered over two evenings on June 9-10, 2022 by Robert Young (Manchester), Jang-Kyo Kim (UNSW) and Luyi Sun (Connecticut) in English as well as Liqun Zhang (BUCT), Yan Li (Tongji) and Ming-Qiu Zhang (SYSU) in Mandarin (Fig. 4.62). They were outstanding lectures as could be evidenced by the extraordinary high attendance albeit virtually.

Fig. 4.59   Symposium held in HKUST in honor of my 70th birthday. First row from left: 5th (Liqun Zhang *CAE*), 6th (Lin Ye *FTSE*), 7th (Huajian Gao *NAS NAE*), 8th (me), 9th (Shangyi Du *CAE*), 10th (Victor Li), 11th (Adrian Mouritz *FTSE*) and 12th (Jingshen Wu).

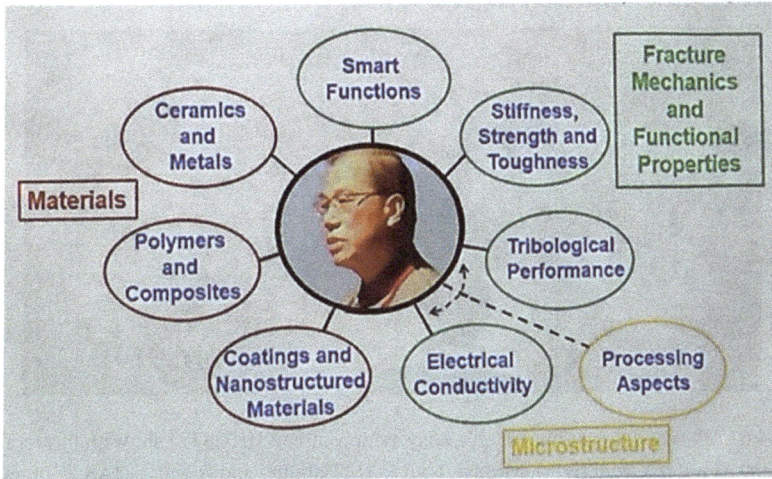

Fig. 4.60   Prof. Klaus Friedrich's description of what my research has covered in my 70th birthday symposium held at HKUST.

I provided my PhD students (supervised and co-supervised no fewer than 65) and postdocs (well over 50) regular opportunities to interact and network with top visiting scientists in the field in addition to accessing the best research training and leadership development at the CAMT. I am very

Fig. 4.61   Photos taken at my 70th birthday symposium at HKUST. Top: With former PhD students (I am 6th from right). Middle: with CAMT alumni and friends at IAS Conference Lodge (I am 4th on the left side of table, Liqun Zang *CAE* is 1st and Ning Hu VP of Hubei University of Technology is 4th on the right side of table). Bottom: at the banquet with Huajian Gao *NAS NAE* (facing camera) and Victor Li (seated first right backing camera).

proud of their achievements. Some are now leaders in their chosen fields in academia, finance and business.

The CAMT alumni and alumnae have a broad geographical base. Before the COVID, it was easy to travel internationally. I could meet up with many

Fig. 4.62   Webinars in honor of my 75th birthday.

Fig. 4.63   (L to R) Yanbin Ke, me, Brian Cotterell and Xiaozhi Hu *ca*. 1987.

of them, especially those in China, and we talked about the good old days (Figs. 4.63–4.65) and their recent research activities. Sometimes this would lead to collaborative research projects, which are often new and challenging to me, and/or joint PhD training programs through bilateral agreements with

Fig. 4.64   Happy times at various meetings. (a) Dinner with CAMT alumni in Beijing. (b) Discussion meeting in HUST (I am sitting on the sofa 2nd from right). (c) With Xiaolin Xie, Jang-Kyo Kim, Klaus Friedrich, Tsu-Wei Chou, Marino Quaresimin and Ming-Qiu Zhang (I am 2nd from right). (d) With CAMT alumni in Chibi (赤壁 Red Cliffs) on a very hot day (I am holding the black umbrella in the middle).

Fig. 4.65   L to R: Dinner with Lin Ye *FTSE*, Jimmy Hsia (then Carnegie Mellon, now NTU Singapore) and Khin Yong Lam (Provost of NTU Singapore) in Sydney Chinatown 2017.

simultaneous supports from the Chinese Scholarship Council. Since Prof. Lin Ye's recent departure to SUSTech in Shenzhen, China, which is a huge loss to us, Prof. Qing Li is now the Director of the CAMT who will lead us to higher achievements in other emerging R&D areas.

4.2.10.2 *My reflections and my moments*

Some reflections

(a) In reminiscing my journeys in research, whatever successes and achievements I made on the way, I owe a lot to my mentors, my colleagues and my research team of graduate students, postdocs and visitors. In my time at the University of Sydney, I had understanding and accommodating Heads of Department and Deans of Engineering who appreciated research excellence and provided the space and freedom for me to flourish and establish my world reputation and international standing in the fields of engineering sciences and composite materials. Chance encounters with different people I mentioned in the different sections above and the opportunities therefrom pointed me to various research areas I had not anticipated. I was well grounded in the basic sciences for my first degree at HKU and that is a necessary condition. Moreover, my wider experiences and exposures to fibre composites, ceramics and polymers as a young postdoc at the University of Michigan and Imperial College London all prepared me well for these new challenges. There is always an element of luck. It is often said that being at the right place at the right time is critical in scientific discovery. I see it no different in engineering research. My work on grain-bridging of coarse-grained alumina ceramics with Brian Lawn at the NIST is a case in point. Same for my other work on asbestos-free fibre cements for James Hardie Industries (right place) with a tight timeline for phasing out asbestos (right time).

(b) When I was a graduate student, there was no pressure to publish and only few authors were on the paper. Today, there is excessive emphasis to publish in journals with high impact factors. I know why but this is not healthy. Charles Gurney did not publish many papers and thought it would suffice only to publish highly original or transformative concepts and leave others to do the follow-up work. Gordon Williams believes only to publish papers in those journals where they are read. Brian Cotterell is even of the view that many academics are just interested to get a grant and publish and repeat the cycle again and again. The usefulness or otherwise of the work is never questioned. For engineering research, I believe we should instill in our next-generation researchers the importance of impacts beyond academia. That is, they should have translational effects on our professions, industry, healthcare and business. We should also not forget our societal responsibility to make this world a better place for all. In this context, I have always valued the work we contributed to the development of asbestos-free fiber cements which certainly has significant health and social-economic benefits for mankind. In addition, our research on the development of the essential work of fracture on polymer thin sheets has helped develop the *ESIS TC-4: Testing Protocol* for *Essential Work* of *Fracture (1993)*. I am most pleased that this has led to an *ISO/Draft International Standard* (ISO/DIS 23524(en) *Plastics — Determination of fracture toughness of films and thin sheets: the essential work of fracture (EWF) method* – ISO/TC 61/SC 2, dated 2022-01) which has been prepared and is now in the approval stage.

(c) I have been closely associated and held formal positions with several professional organizations in my research areas for a long time. These helped me develop strong research networks with some of the very brilliant scientists in their field. I learned from them not be afraid to challenge or debate authorities. I gave the Scala Lecture at the 2017 ICCM-21 meeting in Xi'an (Figs. 4.66 and 4.67). I was the instigator to set up the Asia-Australasian Association for Composite Materials in 1997 in Brisbane, Australia, with the main aims to promote research, education and training, collaboration and application of composites with member countries in the region. We run a series of biennial Asia-Australasian Conference on Composite Materials (AACM). The first AACM-1 was held in Osaka, Japan in 1998 (Fig. 4.68) and I was the inaugural president. Today, ACCM with ECCM and ICCM have become the three major meetings for composites researchers and other stake holders. I was also closely associated with the International Congress on Fracture (ICF) (Fig. 4.69), having been its president (2001-05), won the Takeo Yokobori Gold Medal and presented the Closing Lecture in ICF-13 (2013) in Beijing, China.

(d) Unlike today, 40-50 years ago, there were very few Chinese academics in Australian universities let alone in engineering. My former teacher at HKU, Prof. Yau-Kai Cheung of Finite Element Method and Finite Strip Method fame, was the first Chinese (from Hong Kong) engineering

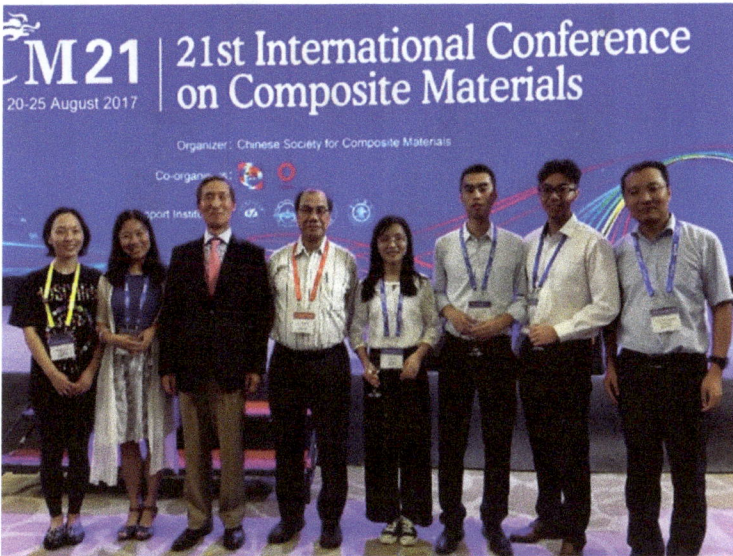

Fig. 4.66   Wth Jang-Kyo Kim *FHKEng* and his team at HKUST at ICCM21 in Xian 2017. I am 4th from the left.

Fig. 4.67   (L-R): At 2017 ICCM21 in Xian with Shanyi Du *CAE*, me, Stephen Tsai *NAE* and Tongyi Zhang *CAS*.

professor appointed (1974-77) at the University of Adelaide. He died on September 23, 2022 after a long illness. I was the second Chinese (also from Hong Kong) engineering professor appointed a decade later in 1987 at the University of Sydney. This situation was not ideal.

Fig. 4.68   with (L) Hiroyuki Hamada and (R) Seeram Ramakrishna *FREng* at ACCM-1, Osaka 1998. I am in the middle.

Fig. 4.69   With Rob Ritchie *FRS NAE* and Brian Lawn *NAE FAA* at ICF10 in Honolulu, Hawaii, 2001. I am on the right.

Therefore, we established a New South Wales Chinese-Australian Academics Society in 1970s to offer a meeting place to discuss academic issues for all Chinese academics working in various universities in the State of NSW. I was an active member and became its president in early 1990s. The Society later changed its name to the Chinese-Australian Academics Society (Fig. 4.70) by dropping NSW to cover all the States and the Territories in the country.

My moments

I am very proud of the Centre for Advanced Materials Technology (CAMT) that I established in 1989 at Sydney and even prouder of the many excellent doctoral students and postdocs I have advised since then. They now have

Fig. 4.70 NSW Australian-Chinese Academic Association with Sydney University Vice-Chancellor Michael Spence (sitting 8th from left) welcoming New Chinese Consul General Shan Hu (sitting 7th from left) to Sydney (2009). I am sitting 6th from right.

their own research groups and the CAMT traditions and values are passing on to future generations. Their hard work and memorable contributions to basic knowledge in our research fields have won us high accolades. I will always be grateful to them for bringing me many happy moments of my life:

- Awarded the AA Griffith Medal in 2016 by the UK Institute of Materials, Minerals and Mining (IOM3). I am honored to follow my PhD supervisor, Prof. Charles Gurney, who was awarded the same medal in 1993.
- Awarded the 2016 AGM Michel Medal of the Mechanical College of Engineers Australia for life-long achievements in mechanical engineering.
- Appointed a Member of the Order of Australia (AM) in 2010 (Fig. 4.71) by the Australian federal government for service to engineering, particularly in the fields of advanced fibre composites and fracture research
- Received an honorary Doctor of Science degree (*honoris causa*) in 2013 (Fig. 4.72) by my *alma mater*, the University of Hong Kong. This recognition made me feeling emotional since I did not expect it.
- Elected in 2008 to the Royal Society of London, which is the world's oldest continuous science academy since 1660, is a special honor for one like me educated in the British Hong Kong colonial system. I never thought I would or could achieve this distinction and signed on the *same* Book of Signatures as Newton, Darwin and Hawking!
- Elected in 2017 as a Foreign Member of the Chinese Academy of Engineering gave me special pride as an ethnic Chinese (Figs. 4.73 and 4.74). I had long collaborated with researchers in many Chinese universities

Fig. 4.71   With NSW Governor (Dame Marie Bashir) in Government House after the AM award ceremony in 2010.

Fig. 4.72   Award of DSc (*honoris causa*) at HKU in 2013 celebrating with family members. (L-R) Man-Ping Luk (sister-in-law), Chung-Sum Mai (sister), Patrick Cheuk & Gina Mai (niece), me, Gabriel Mai (nephew) & Casta Siu, and Yiu-Shang Mai (brother).

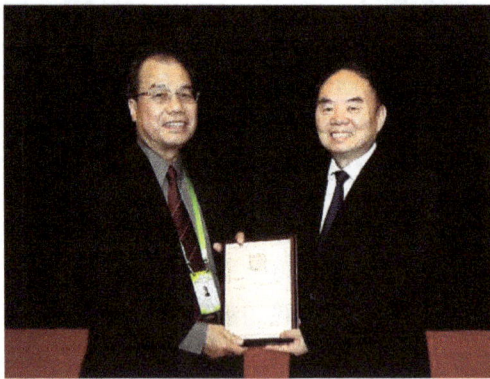

Fig. 4.73   Election to Foreign Member of the Chinese Academy of Engineering in 2017 with Prof. Ji Zhou (President CAE).

Fig. 4.74   Celebration of my election as CAE foreign member organized by the Chinese Consular Education Office in Sydney in 2018. I am sitting 5th from left.

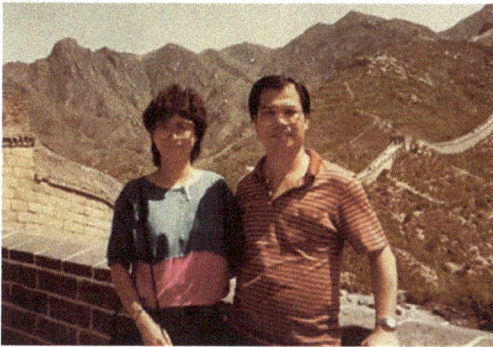

Fig. 4.75   My wife, Lousia, and I stood on the Great Wall in 1983.

since 1979. When my wife and I first stepped on the Great Wall in 1983 and looked at the huge mountainous range far beyond (Fig. 4.75), the flow of emotions and feelings were uncontrolled. The Great Wall is unfortunately no longer the same, now and then.

All earthly glories will pass. All historical figures of times long gone will be washed away with the breaking waves of the East-flowing Yangtze River. I thank my wife, Louisa, for providing me a happy family life and for walking with me, thick and thin, through the intertwining paths of the academic jungle (Fig. 4.76).

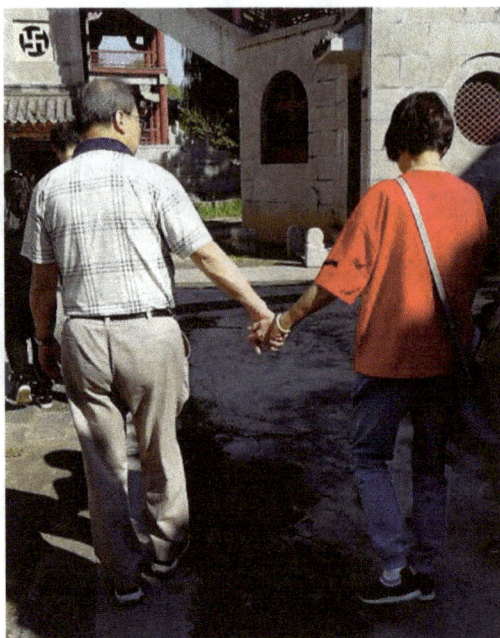

Fig. 4.76　Hand in hand with my wife as we age (執子之手, 與子偕老). Photo taken at Tai Shan (泰山) in 2018.

Index